Education in a Competitive and Globalizing World

Science, Technology, Engineering, and Mathematics Education

Trends and Alignment with Workforce Needs

EDUCATION IN A COMPETITIVE AND GLOBALIZING WORLD

Additional books in this series can be found on Nova's website under the Series tab.

Additional e-books in this series can be found on Nova's website under the e-book tab.

EDUCATION IN A COMPETITIVE AND GLOBALIZING WORLD

SCIENCE, TECHNOLOGY, ENGINEERING, AND MATHEMATICS EDUCATION

TRENDS AND ALIGNMENT WITH WORKFORCE NEEDS

TIMOTHY CURTIS
EDITOR

nova publishers
New York

For permission to use material from this book please contact us:
Telephone 631-231-7269; Fax 631-231-8175
Web Site: http://www.novapublishers.com

Additional color graphics may be available in the e-book version of this book.

Library of Congress Cataloging-in-Publication Data

ISBN: 978-1-63463-126-6

Published by Nova Science Publishers, Inc. † New York

CONTENTS

PREFACE

Chapter 1 - The term “STEM education” refers to teaching and learning in the fields of science, technology, engineering, and mathematics. It typically includes educational activities across all grade levels— from pre-school to post-doctorate—in both formal (e.g., classrooms) and informal (e.g., afterschool programs) settings. Federal policymakers have an active and enduring interest in STEM education and the topic is frequently raised in federal science, education, workforce, national security, and immigration policy debates. For example, more than 225 bills containing the term “science education” were introduced between the 102th and 112th congresses.

The United States is widely believed to perform poorly in STEM education. However, the data paint a complicated picture. By some measures, U.S. students appear to be doing quite well. For example, overall graduate enrollments in science and engineering (S&E) grew 35% over the last decade. Further, S&E enrollments for Hispanic/Latino, American Indian/Alaska Native, and African American students (all of whom are generally underrepresented in S&E) grew by 65%, 55%, and 50%, respectively. On the other hand, concerns remain about persistent academic achievement gaps between various demographic groups, STEM teacher quality, the rankings of U.S. students on international STEM assessments, foreign student enrollments and increased education attainment in other countries, and the ability of the U.S. STEM education system to meet domestic demand for STEM labor.

Various attempts to assess the federal STEM education effort have produced different estimates of its scope and scale. Analysts have identified between 105 and 252 STEM education programs or activities at 13 to 15 federal agencies. Annual federal appropriations for STEM education are typically in the range of $2.8 billion to $3.4 billion. All published inventories identify the Department of Education, National Science Foundation, and Health and Human Services as key agencies in the federal effort. Over half of federal STEM education funding is intended to serve the needs of postsecondary schools and students; the remainder goes to efforts at the kindergarten-through-Grade 12 level. Much of the funding for post-secondary students is in the form of financial aid.

Federal STEM education policy concerns center on issues that relate to STEM education as a whole—such as governance of the federal effort and broadening participation of underrepresented populations—as well as those that are specific to STEM education at the elementary, secondary, and postsecondary levels. Governance concerns focus on perceived duplication and lack of coordination in the federal effort; broadening participation concerns tend to highlight achievement gaps between various demographic groups. Analysts suggest a variety of policy proposals in elementary, secondary, and postsecondary STEM education. At the K-12 level, these include proposals to address teacher quality, accountability, and standards. At the post-secondary level, proposals center on efforts to remediate and retain students in STEM majors.

This report is intended to serve as a primer for outlining existing STEM education policy issues and programs. It includes assessments of the federal STEM education effort and the condition of STEM education in the United States, as well as an analysis of several of the policy issues central to the contemporary federal conversation about STEM education. Appendix A contains frequently cited data and sources and Appendix B includes a selection of major STEM-related acts.

Chapter 2 - Federal STEM education programs help enhance the nation's global competitiveness by preparing students for STEM careers. Researchers disagree about whether there are enough STEM workers to meet employer demand. GAO was asked to study the extent to which STEM education programs are aligned with workforce needs.

GAO examined (1) recent trends in the number of degrees and jobs in STEM fields, (2) the extent to which federal postsecondary STEM education programs take workforce needs into consideration, and (3) the extent to which federal K-12 STEM education programs prepare students for postsecondary STEM education. GAO analyzed trends in STEM degrees and jobs since 2002 using 3 data sets—the Integrated Postsecondary Education Data System, American Community Survey, and Occupational Employment Statistics—and surveyed 158 federal STEM education programs. There were 154 survey respondents (97 percent): 124 postsecondary and 30 K-12 programs. In addition, GAO conducted in-depth reviews—including interviews with federal officials and grantees—of 13 programs chosen from among those with the highest reported obligations.

In: Science, Technology, Engineering, … ISBN: 978-1-63463-126-6
Editor: Timothy Curtis

Chapter 1

SCIENCE, TECHNOLOGY, ENGINEERING, AND MATHEMATICS (STEM) EDUCATION: A PRIMER*

Heather B. Gonzalez and Jeffrey J. Kuenzi

SUMMARY

The term "STEM education" refers to teaching and learning in the fields of science, technology, engineering, and mathematics. It typically includes educational activities across all grade levels— from pre-school to post-doctorate—in both formal (e.g., classrooms) and informal (e.g., afterschool programs) settings. Federal policymakers have an active and enduring interest in STEM education and the topic is frequently raised in federal science, education, workforce, national security, and immigration policy debates. For example, more than 225 bills containing the term "science education" were introduced between the 102th and 112th congresses.

The United States is widely believed to perform poorly in STEM education. However, the data paint a complicated picture. By some measures, U.S. students appear to be doing quite well. For example, overall graduate enrollments in science and engineering (S&E) grew 35% over the last decade. Further, S&E enrollments for Hispanic/Latino, American Indian/Alaska Native, and African American students (all of

* This is an edited, reformatted and augmented version of Congressional Research Service, Publication No. R42642, dated April 5, 2013.

whom are generally underrepresented in S&E) grew by 65%, 55%, and 50%, respectively. On the other hand, concerns remain about persistent academic achievement gaps between various demographic groups, STEM teacher quality, the rankings of U.S. students on international STEM assessments, foreign student enrollments and increased education attainment in other countries, and the ability of the U.S. STEM education system to meet domestic demand for STEM labor.

Various attempts to assess the federal STEM education effort have produced different estimates of its scope and scale. Analysts have identified between 105 and 252 STEM education programs or activities at 13 to 15 federal agencies. Annual federal appropriations for STEM education are typically in the range of $2.8 billion to $3.4 billion. All published inventories identify the Department of Education, National Science Foundation, and Health and Human Services as key agencies in the federal effort. Over half of federal STEM education funding is intended to serve the needs of postsecondary schools and students; the remainder goes to efforts at the kindergarten-through-Grade 12 level. Much of the funding for post-secondary students is in the form of financial aid.

Federal STEM education policy concerns center on issues that relate to STEM education as a whole—such as governance of the federal effort and broadening participation of underrepresented populations—as well as those that are specific to STEM education at the elementary, secondary, and postsecondary levels. Governance concerns focus on perceived duplication and lack of coordination in the federal effort; broadening participation concerns tend to highlight achievement gaps between various demographic groups. Analysts suggest a variety of policy proposals in elementary, secondary, and postsecondary STEM education. At the K-12 level, these include proposals to address teacher quality, accountability, and standards. At the post-secondary level, proposals center on efforts to remediate and retain students in STEM majors.

This report is intended to serve as a primer for outlining existing STEM education policy issues and programs. It includes assessments of the federal STEM education effort and the condition of STEM education in the United States, as well as an analysis of several of the policy issues central to the contemporary federal conversation about STEM education. Appendix A contains frequently cited data and sources and Appendix B includes a selection of major STEM-related acts.

INTRODUCTION

The term "STEM education" refers to teaching and learning in the fields of science, technology, engineering, and mathematics. It typically includes educational activities across all grade levels— from pre-school to post-doctorate—in both formal (e.g., classrooms) and informal (e.g., afterschool programs) settings. Federal policymakers have an active and enduring interest in STEM education and the topic is frequently raised in federal science, education, workforce, national security, and immigration policy debates. The purpose of this report is to put various legislative and executive branch STEM education-related policy proposals into a useful context.

Although many observers cite the launch of the Soviet Union's Sputnik satellite in the 1950s as a key turning point for STEM education policy in the United States, federal interest in scientific and technological literacy writ large is longstanding and dates to at least the first Congress.[1] For example, in the first State of the Union address President George Washington called upon Congress to promote scientific knowledge for the sake of the republic and the polity, saying

> Nor am I less persuaded that you will agree with me in opinion that there is nothing which can better deserve your patronage than the promotion of science and literature. Knowledge is in every country the surest basis of public happiness. In one in which the measures of government receive their impressions so immediately from the sense of the community as in ours it is proportionably [sic] essential.[2]

More recent concerns about scientific and technological literacy in the United States focus on the relationship between STEM education and national prosperity and power. Since World War II, the United States has benefitted from economic and military advances made possible, in part, by a highly skilled STEM workforce. However, today the economic and social benefits of scientific thinking and STEM education are widely believed to have broad application for workers in both STEM and non-STEM occupations.[3] As such, many contemporary policymakers consider widespread STEM literacy, as well as specific STEM expertise, to be critical human capital competencies for a 21st century economy.[4]

The primary domestic source of STEM labor in the United States is the education system.[5] Federal legislators have paid close attention the STEM-related outputs of that system—such as the number of college graduates with degrees in STEM fields or the performance of U.S. students on international mathematics tests—and have sought to increase its functioning and capacity though federal policy and investments. For example, over 225 bills containing the term "science education" were introduced in the 20 years between the 102nd (1991-1992) and 112th (2011-2012) congresses. Agency reauthorization bills often contain STEM education-related provisions and at least 13 federal agencies conduct STEM education programs or activities. The federal investment in STEM education programs is estimated at between $2.8 billion and $3.4 billion annually.[6]

Congressional interest in STEM education heightened in 2007 when the National Academies published a report titled *Rising Above the Gathering Storm: Energizing and Employing America for a Brighter Economic Future.*[10] This influential publication warned federal policymakers that perceived weaknesses in the existing U.S. STEM education system—along with other important factors—threatened national prosperity and power. Although some analysts disputed its assertions, the report helped focus the federal conversation about STEM education and led, in part, to passage of the America Creating Opportunities to Meaningfully Promote Excellence in Technology, Education, and Science Act (or America COMPETES Act). Among other things, that act authorized STEM education programs at the National Science Foundation (NSF), National Oceanic and Atmospheric Administration (NOAA), Department of Energy (DOE), and Department of Education (ED).

What Is STEM?

Whether it is visas for foreign workers, scholarships for STEM majors, or funding for scientific research, the question of what we mean by the term "STEM" is central to the federal policy conversation. Some federal agencies, such as the NSF, use a broader definition of STEM that includes psychology and the social sciences (e.g., political science, economics) as well as the so-called core sciences and engineering (e.g., physics, chemistry, mathematics).[7] Others, including the Department of Homeland Security (DHS), U.S. Immigration and Customs Enforcement (ICE), use a narrower definition that generally excludes social sciences and focuses on mathematics, chemistry, physics, computer and information sciences, and engineering.[8] Some analysts argue that field-specific definitions such as these are too static and that definitions of STEM should focus on "an assemblage of practices and processes that transcend disciplinary lines and from which knowledge and learning of a particular kind emerges."[9]

Congress reauthorized the America COMPETES Act in 2010 (P.L. 111-358), thereby advancing it to the implementation phase of the policy cycle.

Congress may opt to reauthorize the act in 2013, when many of its provisions will expire. In the meantime the federal conversation about STEM education continues in the budget and appropriations processes and in the various STEM education-related bills introduced each year.

Given policymakers' ongoing interest in establishing the scope and scale of federal STEM education effort, the first section of this report examines federal agencies, programs, and funding for STEM education. The second section examines the performance of the U.S. STEM education system and includes data and sources frequently cited in federal STEM education policy debates. The third section analyzes various STEM education policy issues and options, including those that relate to STEM education as a whole and those that are specific to the kindergarten-throughgrade-12 (K-12) and higher education systems. **Appendix A** and **Appendix B** contain links to sources of STEM education data and publications and to selected major legislation in federal STEM education policy history.

THE FEDERAL EFFORT IN STEM EDUCATION

At the request of Congress, four inventories of federal STEM education programs and activities have been published in recent years; two by the Government Accountability Office (GAO), one by the Academic Competitiveness Council (ACC),[11] and one by the National Science and Technology Council (NSTC).[12] The first GAO study, in 2005, found 207 distinct federal STEM education programs funded at about $2.8 billion in FY2004 (hereinafter this report is referred to as "GAO-2005").[13] In 2007, the ACC found 105 STEM education programs funded at about $3.1 billion in FY2006 (hereinafter this report is referred to as "ACC-2007").[14] A 2011 report by the NSTC identified 252 "distinct investments" in STEM education funded at about $3.4 billion in FY2010 (hereinafter this report is referred to as "NSTC-2011").[15] A second GAO study, published in 2012, reported 209 programs funded at about $3.1 billion in FY2010 (hereinafter this report is referred to as "GAO-2012").[16]

The discrepancies between these inventories indicate that establishing the federal effort in STEM education is complex and subject to methodological challenges. Differences between the inventories are due, in part, to the lack of a common definition of what constitutes a STEM education program or activity.

Auditors have also found STEM education activities performed by science mission agencies difficult to capture because such activities tend to be fiscally and organizationally integrated into what are otherwise primarily scientific programs.

Funding calculations and program identification become even more intricate when broad-purpose education programs with a STEM goal are considered (e.g., teacher training programs that focus on mathematics in addition to other fields such as reading). Finally, some estimates of federal STEM education activities depend on agency self-reporting, which is a less-reliable auditing method.

Despite these limitations, these four inventories reveal several general patterns in federal STEM education investments. The next two sections will discuss the inventories' findings by federal agency, population served, and program objective.

Federal Programs by Agency

Each of the four congressionally mandated inventories of the federal STEM education effort found that virtually all federal agencies administer STEM education programs. However, three agencies account for about four-fifths of federal funding for STEM education: the National Science Foundation (NSF) and the Departments of Education (ED) and Health and Human Services (HHS).

As **Figure 1** shows, all four inventories found that about one-third of the federal investment in STEM education is appropriated to the NSF.

However, each inventory found different portions of STEM education funding at the other two agencies. The GAO-2005 and ACC-2007 inventories found a larger share of STEM funding at HHS than the GAO-2012 and NSTC-2011 studies. The GAO-2005 inventory found a much smaller share of funds at ED (8%; compared to 23%, 29%, and 22% in the latter three inventories). This discrepancy is likely attributable to a large increase in the FY2006 ED appropriation (roughly $310 million) for the National Science and Mathematics Access to Retain Talent (SMART) Grant program, which was newly authorized in 2005. Authority for the SMART Grant program ended in FY2010. No funds have been appropriated for the program since then.[17]

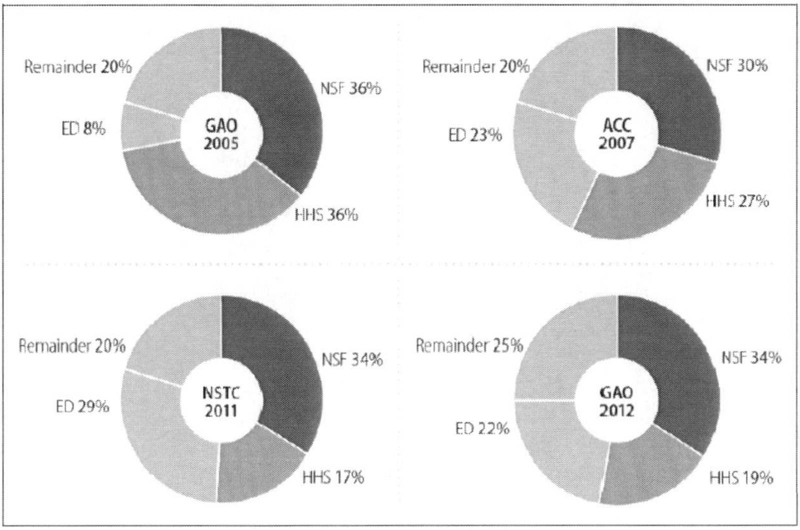

Source: CRS calculation based on GAO-2005, Figure 1; ACC-2007, Page 21; NSTC-2011, Figure 11; and GAO-2012, Appendix 2.

Figure 1. Federal STEM Education Funding, by Agency.

Population Served and Program Objectives

Each inventory took a different methodological approach and reported results somewhat incompatibly in terms of population served and program objective. This incompatibility is likely due to overlap between the populations served or program objectives within the individual STEM education programs. That is to say, sometimes the same program serves multiple populations (e.g., high school students and postsecondary students, graduate students and postdoctoral fellows). Additionally, nearly all STEM education programs have multiple objectives (e.g., supporting research and increasing degree attainment, encouraging advanced study and smoothing career transitions). Further, the inventories reported their findings on populations served and program objectives in different ways, thus making it difficult to compare their results on these important program elements.

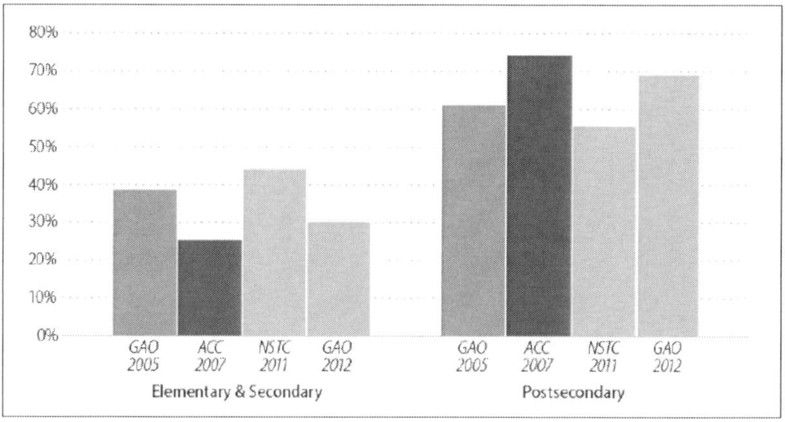

Source: CRS calculation based on GAO-2005, Table 8; ACC-2007, Page 2; NSTC-2011, Table 6; GAO-2012, Page 15.

Figure 2. Percentage of STEM Education Programs, by Education Level.

Each inventory reported on programs by population served (e.g., by education level), although both GAO studies did this only in terms of the number of programs and not their funding level. Figure 2 presents the percentage of programs primarily serving elementary and secondary schools and postsecondary institutions as a share of the total number of programs identified in each inventory. Of the programs identified by GAO-2005, just fewer than 40% served elementary and secondary schools and just over 60% served postsecondary institutions; compared to 25% and 75% in the ACC-2007 study, 44% and 56% in the NSTC-2011 inventory, and 31% and 69% in the GAO-2012 study. The NSTC-2011 inventory did not include programs serving postdoctoral fellows, thus lowering the share of programs found at the postsecondary level.

Each inventory also collected information on program objectives. However, only the NSTC-2011 and GAO-2005 inventories reported information that could be summarized. The NSTC-2011 study found that nearly three-quarters of programs (74%) have at least two secondary program objectives in addition to the primary objective.[18] The GAO-2005 study found an even larger share of programs (80%) with more than one goal, with about half supporting four or more goals.[19]

Figure 3 presents the share of federal STEM education programs by primary program objective for the GAO-2005 and NSTC-2011 inventories. In both studies, the majority of programs support degree attainment, research experience, and career development for postsecondary students (57% in the GAO study and 59% in the NSTC study). Fewer than one in five programs support STEM learning and engagement (GAO, 18%; NSTC, 13%). About one in ten programs support the training of STEM educators (GAO, 11%; NSTC, 9%).

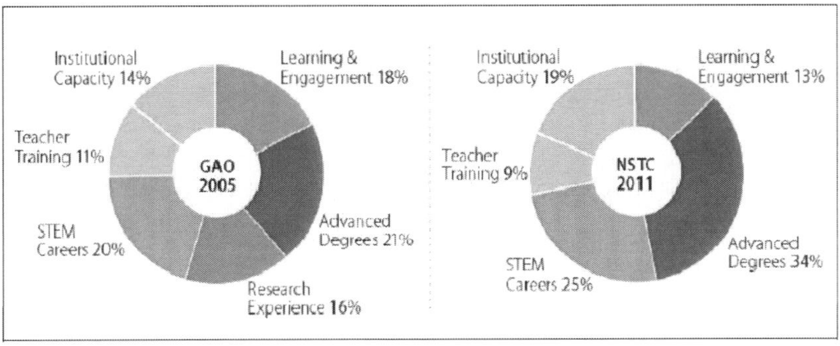

Source: CRS calculation based on GAO-2005, Table 6; NSTC-2011, Figure 7.

Figure 3. Percentage of STEM Education Programs, by Primary Objective.

Selected STEM Education Programs

In FY2012, the largest federal programs supporting STEM education were the Ruth L. Kirschstein National Research Service Awards ($274 million)[20] administered by HHS, the Graduate Research Fellowships program ($198 million) administered by NSF, and the Mathematics and Science Partnership program ($150 million) administered by ED. Not only are these the largest programs, they also represent two of the major activities receiving federal support, namely fellowships for graduate study and K-12 teacher training.

Ruth L. Kirschstein National Research Service Awards (HHS)

First funded in 1975, the Ruth L. Kirschstein National Research Service Awards (NRSA) constitute just under half (roughly 48%) of HHS spending on STEM education.[21] Most NRSA funds support Institutional Research Training Grants. About 15%-20% of funds support individual fellowships. The Institutional Research Training Grants are awarded to institutions to develop or enhance research training opportunities for individuals, selected by the institution, who are training for careers in specified areas of interest to the institution or principal investigator.[22] The individual fellowships are awarded directly to individuals from various organizations within the National Institutes of Health (e.g., the National Institute on Aging) to support the particular research interests of the individual receiving the award.[23]

Graduate Research Fellowships (GRF)

The Graduate Research Fellowship (GRF) program is the oldest and largest STEM education program at NSF. Established in 1952, the GRF is one of the most prestigious national awards offered to STEM graduate students. Fellows receive three-year portable stipends for graduate study leading to research-based master's and doctoral degrees in fields related to NSF's mission. Applicants are chosen by merit review. The NSF issued 7,800 fellowships (including 2,000 new fellowships) worth up to $42,000 each in FY2012. This amount includes a $12,000 cost-ofeducation (COE) allowance for the enrolling institution and a $30,000 stipend for the fellow.[24]

Some of the policy concerns associated with the GRF program focus on the number of fellowships offered annually, stipend and COE levels, and source(s) of funding.[25] Historically, funding for the GRF came primarily from NSF's main education account. Section 2 of the America COMPETES Reauthorization Act of 2010 (P.L. 111-358) directed NSF to fund the GRF equally from both the main education *and* research accounts. Funding for the GRF program increased after this change.[26] Some analysts propose expanding the GRF by creating a new NSF-industry fellows program, the funding for which would come equally from the private and public sectors.[27]

Mathematics and Science Partnerships (MSP)

The Mathematics and Science Partnership (MSP) program accounted for more than half (52%) of ED's STEM education portfolio in FY2012 ($150 million of $284 million). First authorized by the No Child Left Behind Act of 2001 (P.L. 107-110), the MSP program provides formula grants to states to increase the academic achievement of students in mathematics and science by enhancing the content knowledge and teaching skills of classroom teachers. With these funds, each State administers a grant competition in which awards are made to partnerships between high-need school districts and institutions of higher education. Grantees typically provide summer institutes and ongoing professional development designed to improve teachers' content knowledge through direct collaboration with scientists, mathematicians, and engineers.

In addition to ED's MSP, the 107th Congress created a companion program through the National Science Foundation Authorization Act of 2002 (P.L. 107-368). NSF's companion program is also called the Mathematics and

Science Partnership (NSF-MSP) program.[28] Funded at $55 million in FY2012, NSF-MSP is a research and development effort that supports projects to serve as models of innovation for K-12 STEM education through competitive grants to institutions of higher education or nonprofit organizations in partnership with local education agencies. The NSF Director is required to report annually to Congress on how the program has been coordinated with ED's MSP program.[29]

THE CONDITION OF U.S. STEM EDUCATION

No single fact or statistic can wholly capture the condition of STEM education in the United States and for a variety of reasons the question "what is the condition of STEM education in the United States?" may be unanswerable. However, some trends appear to have held over time and in the most general sense, the condition of STEM education in the United States may be characterized as having more or less held constant or improved over the course of the last four decades. This is not the end of the story though. Looking at STEM education from this broad perspective disguises trends that concern many analysts and drive policy in this area. Among these concerns are persistent achievement gaps between various demographic groups, U.S. student performance on international mathematics and science tests, foreign student enrollments in U.S. institutions of higher education, global STEM education attainment, U.S. STEM teacher quality, and the U.S. STEM labor supply.

Upward Trends

According to the U.S. Census Bureau, the percentage of U.S. bachelor's degree holders with undergraduate degrees in science and engineering (S&E) was 36.4% in 2009 (approximately 20 million people).[30] This percentage is roughly consistent with the annual domestic production of S&E bachelor's degrees. The NSF estimates that the percentage of bachelor's degrees in S&E fields has held relatively constant—at between approximately 30% and 35% of all bachelor's degrees—for the past four decades. However, because the U.S. college-age population grew during these years, the total number of S&E bachelor's degrees awarded annually more than doubled between 1966 and 2008 (from 184,313 to 494,627).[31]

At the graduate level, S&E degrees predominate doctorate degree production. Since 1966, the percentage of doctorates in S&E fields has ranged between approximately 56% and 67% of all graduate degrees (where a field of study has been reported).[32] The total number of doctoral degrees in S&E fields has nearly tripled, growing from 11,570 in 1966 to 32,827 in 2008.[33] Graduate enrollments show similar upward trends. In 2010 there were 556,532 graduates enrolled in S&E fields (an historic peak), up from 413,536 in 2000.[34] Figure 4 displays the number of S&E degrees awarded, by degree level, over the last four decades.

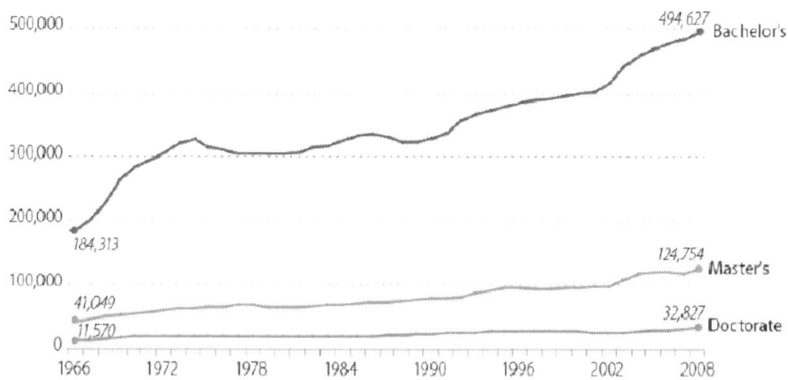

Source: National Science Foundation, National Center for Science and Engineering Statistics, "Table 1. S&E Degrees 1966-2008," *Detailed Statistical Tables* (NSF 11-316).

Notes: Includes only degrees where field of study is known. Includes degrees awarded in the social sciences and psychology.

Figure 4. Number of S&E Degrees Awarded from 1966-2008, By Degree Level.

Similar consistency in performance over time may also be found in the lower grades. The performance of U.S. K-12 students on standardized national mathematics tests has held constant or improved over the past four decades. For example, the average National Assessment of Educational Progress (NAEP) mathematics scores of students in 4th and 8th grades, grades in which students have been tested for decades, increased by 28 and 21 points—respectively— between 1990 and 2011.[35] **Figure 5** presents average NAEP math scores by various student subpopulations. Although all group scores have improved over time, sizable gaps remain.

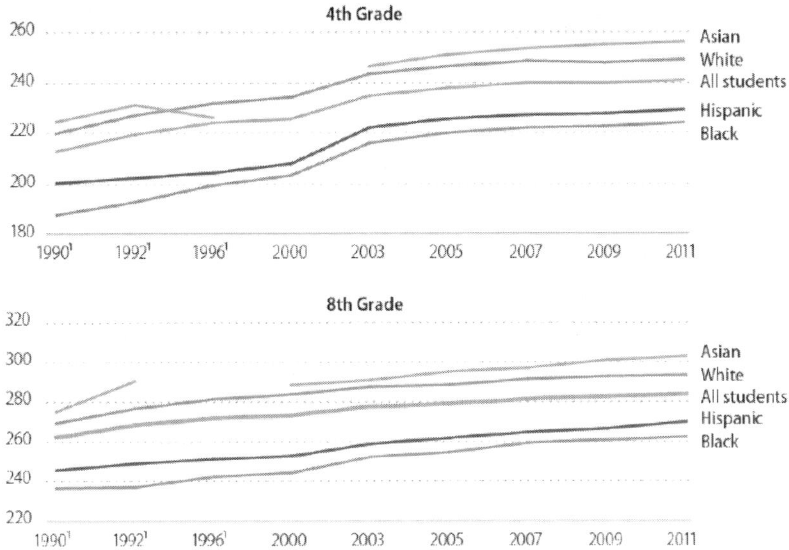

Source: CRS analysis of data from U.S. Department of Education, Institute of Education Sciences, National Center for Education Statistics, National Assessment of Educational Progress, various years.

Notes: The NAEP Mathematics scale ranges from 0 to 500. Some apparent differences between estimates may not be statistically significant. Time series are broken for years in which sample size was insufficient.

[1]Accommodations for students with disabilities were not permitted prior to 1996.

Figure 5. Trends in 4th and 8th Grade Average Mathematics Scores; Main NAEP, 1990 to 2011.

The average scores of 12th grade students on the main NAEP mathematics assessment were three points higher in 2009 than they were in 2005, when the test was first administered to this age group.[36]

Areas of Concern

In some respects, the overall trends paint a fairly optimistic picture for STEM education in the United States. Why, then, are so many observers so concerned about it? Analysts with concerns about STEM education cite a variety of data and trends as alarming. Among these are persistent achievement gaps between various demographic groups, U.S. student performance on international mathematics and science tests, foreign student enrollments in U.S. institutions of higher education, global STEM education attainment, U.S. STEM teacher quality, and the U.S. STEM labor supply.

Academic Achievement Gaps

A central topic in the conversation about STEM education focuses on so-called achievement gaps among various racial and ethnic groups and between women and men in certain STEM education outcomes. These gaps can be seen in a wide variety of STEM data, which show disparities by race, ethnicity, and gender in test scores, degree attainment, and employment. For example, there was at least a 20-point gap between the average scores of white students and their black and Hispanic counterparts on the 2011 4th and 8th grade NAEP mathematics assessments.[37] At the higher education level, only 18.5% of bachelor's degrees in engineering went to women in 2008.[38] Some STEM achievement gaps appear to hold relatively constant over time.

Although achievement gaps appear to be both pervasive and persistent, some evidence points to various types of improvement over time and in certain fields. For example, in the decade between 2000 and 2010, graduate enrollments in S&E fields grew by 35%. Further, among U.S. citizens and permanent residents, S&E graduate enrollments among Hispanic/Latino, American Indian/Alaska Native, and black/African America students grew at a higher rate than that of whites (not of Hispanic origin) and Asians.[39] While women account for relatively small percentages of degree recipients in certain STEM fields (as noted above), they accounted for 77.1% of the psychology degrees and 58.3% of the biological and agricultural sciences degrees in 2008.[40] Finally, although the 20+ point gap between the average scores of white students and their black and Hispanic counterparts on both the 4th and 8th grade NAEP mathematics tests has persisted for two decades, students of all ethnicities and races have higher average scores in 2011 than they did in 1990.[41]

Teacher Quality

Many observers look to the nation's teaching force as a source of shortcomings in student mathematics and science achievement. Research on teacher quality conducted over the last 20 years reveals that, among those who teach mathematics and science, having a major in the subject positively affects student achievement.[42] Unfortunately, many U.S. mathematics and science teachers lack this credential. For example, nearly all high school teachers have at least a baccalaureate degree; however, mathematics teachers are less likely than teachers of other subject areas to have majored in the subject they teach.

In the 2007-2008 school year, roughly 17% of all high school teachers did not major in the subject they taught, while 28% of mathematics teachers did not major in mathematics.[43] Moreover, among those who majored in the subject they taught, mathematics teachers are less likely to be subject-certified than other teachers.

International Assessment Rankings

Another area often of concern is how U.S. students compare with their peers in other nations in their knowledge of mathematics and science. While U.S. students usually outscore the all-country average on international mathematics and science tests, they typically score below the average of industrialized nations. For example, U.S. 15 year-olds ranked below the Organization for Economic Cooperation and Development (OECD) average in mathematics—and ranked at the OECD average in science—on the 2009 Program for International Student Assessment (PISA).[44] U.S. students fare better on the Trends in International Mathematics and Science Study (TIMSS); U.S. 8th graders ranked 9th in mathematics and 11th in science on the 2007 TIMSS assessment.[45]

Many observers caution against using student assessments to compare nations. A variety of factors may influence test results, including translation issues, differences in test administration, student effort,[46] and the selection and diversity of test takers. The latter issue is often raised by critics of international assessments when looking at U.S. student performance. Some observers say that low performance in the United States is closely related to poverty, though the same reasoning applies to other countries. One analysis of the 2009 PISA results found that the richest U.S. areas (especially areas with less than 10% poverty) perform better than most other nations.[47]

Foreign Student Enrollment[48]

Although the number of degrees awarded in STEM fields has increased over time, many analysts are concerned about the percentage of STEM degrees that go to foreign students. For example, foreign students earn roughly one-third of all U.S. S&E doctoral degrees and earn half (or more) of U.S. doctoral degrees in the specific fields of engineering, physics, computer sciences, and economics.[49] Further, the percentage of doctoral degrees going to foreign students has been more or less increasing since the mid-1970s.[50]

The presence of foreign students in U.S. graduate S&E programs has been and continues to be of concern to some analysts because foreign graduates may not be eligible for work in the United States or for certain jobs requiring security clearance. Other observers suggest that these trends may mean missed opportunities or depressed wages for U.S. citizens and permanent residents who may be displaced by foreign graduates. Other analysts say that federal policymakers should encourage foreign STEM students to study and stay in the United States, arguing that policies meant to attract the world's best and brightest are key to ensuring U.S. competitiveness.[51]

Global STEM Education Attainment

In addition to concerns about foreign students obtaining STEM graduate degrees at U.S. institutions, some observers assert that the United States is falling behind other nations in the production of total STEM degrees. Of the 5 million first university degrees (e.g., undergraduate degrees) awarded globally in S&E in 2008, students in China earned about 23%, European Union students earned about 19%, and U.S. students earned about 10%. Further, while the United States awarded the largest number of total S&E doctoral degrees in 2008 (about 33,000), in 2007 China overtook the United States to become the world leader in the number of doctoral degrees awarded in the specific fields of natural sciences and engineering.[52]

Some analysts challenge these degree production numbers, arguing that at least in some cases (e.g., engineering) the United States produces higher quality graduates and that country-level comparisons are misleading because the statistics are not based on common methodologies or definitions.[53] However, attention to degree attainment trends has been amplified by scale differences between the sizes of the United States' and Chinese and Indian populations (i.e., about 300 million in the United States compared to about a 1.34 billion in China and 1.22 billion in India).

U.S. STEM Labor Supply[54]

A primary rationale behind federal STEM education policies relies on their perceived impact on the U.S. S&E workforce—and through it, on U.S. economic competitiveness and national security. Many business, academic, and policy leaders assert that U.S. STEM education weaknesses have contributed (or will soon contribute) to national S&E workforce shortages and that this labor supply problem has diminished U.S. global economic competitiveness and threatened national security (or will do so in the future).[55] However, some analysts argue that perceived limitations in the U.S. S&E workforce are overstated and that U.S. competitiveness is not threatened by across-the-board S&E labor shortages and does not require a supply-side response.[56] A third view holds that perceptions of S&E workforce shortages are accurate if the increasing numbers of jobs that are technically non-STEM, but that require STEM competencies (e.g., analytical skills), are included in labor demand calculations.[57]

Data and Methodological Limitations

Data are a big part of the current STEM education policy debate. Those who advocate for or against various STEM education policy proposals cite a variety of data and statistics in support of their assertions. However, in some cases data showing the impact of policy changes may lag behind those changes by years or decades, making accurate evaluation and policy assessment difficult. In other cases, data may be interpreted or used in ways that do not reflect potentially important research or methodological limitations. For example, one 2010 editorial stated that "the World Economic Forum ranked [the United States] 48th out of 133 developed and developing nations in quality of mathematics and science instruction."[58] The editorial did not explain that the source of the 48th place ranking was an opinion survey of global business executives. Although opinion surveys are often relevant in policy debate, policymakers may interpret their results differently than they would other kinds of evidence. These and other data limitations may challenge federal policymaking in this area.

STEM Education Policy Issues

Stakeholders with an interest in improving STEM education suggest a wide and disparate set of policy options for Congress. Some of these recommendations address governance concerns about the administration of federal programs—e.g., removing duplication and improving program coordination within and across agencies. Other policy options focus attention on elements of the elementary and secondary school system—e.g., improving the quality and quantity of mathematics and science teachers and strengthening school accountability measures. Additional recommendations look to make improvements at the post-secondary level—e.g., enhancing retention of undergraduate STEM majors and strengthening incentives to pursue advanced STEM education. Many options focus on improving the STEM education outcomes of underrepresented populations.

Governance Concerns

Governance concerns are central to the contemporary debate about the federal STEM education effort. The scope and scale of the federal STEM education portfolio has some analysts concerned that federal agencies may be duplicating effort. In response to these concerns, some policymakers have proposed consolidating or eliminating STEM education programs. Other stakeholders support the broad and diffused nature of the federal STEM education effort and are more concerned with an apparent lack of coordination. Proponents of this view have argued for the development of an overarching federal STEM education strategy.

Duplication and Consolidation

Program consolidation is a widely debated policy option for federal STEM education programs. Advocates for this approach perceive duplication in the federal effort and assert that merging programs would result in cost savings. Proposals to consolidate STEM education programs have been made by both members of Congress and the Administration.[59] Some policymakers see program consolidation as a means to increase program flexibility and responsiveness because (under certain models of consolidation) federal program managers would have greater authority to shift priorities.[60] However, other policymakers may object to this change because it typically transfers program control from the legislative to the executive branch, shifting the balance of power.

Consolidation (particularly in the form of block grants) has also been proposed as a strategy to transfer control to the states and as a means to reduce program costs. Shifting control to the states could increase their ability to respond to local conditions and needs, but might make it more difficult to drive a national STEM education agenda[61] or to leverage unique assets of federal science agencies.[62] On the question of cost, the GAO has found that program consolidation can be more expensive in the short term and may not result in long term savings (if workloads are not also reduced) because administrative costs in federal STEM education programs tend to be less than 10% of total program costs.[63] Consolidation opponents raise general concerns about the potential impact of merging programs, arguing that certain programs (such as STEM education programs) need specified funding streams to avoid being passed over in favor of competing education priorities. It is unclear if this assertion would hold true in practice.

The impact of program consolidation on the federal STEM education effort would depend on what programs are consolidated, how the consolidation is accomplished, how funding streams are affected, and the degree to which programs are duplicative.[64] Congress could, for example, seek either a full or a partial consolidation of STEM education programs at individual agencies or across the entire portfolio. If Congress both consolidates programs and reduces funding levels, it may achieve savings from both administrative and program costs. Savings and program impacts from these changes would vary, depending on which of these strategies policymakers pursue.

Duplication or Overlap?

Published assessments of duplication in the federal STEM education portfolio are somewhat contradictory. Preliminary findings from April 2011 GAO testimony appeared to suggest the potential for duplication in federal teacher quality programs, including some STEM teacher programs.[65] However, the NSTC-2011 inventory specifically examined the duplication question within the federal STEM education portfolio and found "little overlap and no duplication."[66] The GAO-2012 inventory concluded that 83% of federal STEM education programs overlapped "to some degree," but stated that this overlap would "not necessarily be duplicative."[67]

Coordination and Strategy

Some stakeholders maintain that duplication in the federal portfolio is limited. They tend to focus instead on a perceived lack of coordination among and within agencies. To address this concern, some analysts call for an overarching STEM education strategy. Until recently, the federal STEM education effort was primarily undertaken in a distributed fashion that responded to the specific needs of agencies and STEM constituencies. That is, in general, programs were not part of a defined government-wide system with clear roles played by individual federal agencies. Some view this distributed approach as particularly sensitive to the unique workforce needs or STEM education assets of federal science agencies; other observers suggest that an overarching strategy may improve the efficiency of federal STEM education investments.[68]

Both the Congress and the President have moved to develop a federal STEM education strategy. Section 101 of the America COMPETES

Reauthorization Act of 2010 (P.L. 111-358) directed the NSTC to develop and implement a five-year federal STEM education strategy. Although the NSTC had not published this strategy by mid-November 2012, it issued a status report in February 2012.[69] That status report identifies two common federal STEM education agency goals—STEM workforce development and STEM literacy—as well as policy and administrative strategies designed to accomplish these goals. In particular, the status report identifies four priority policy areas for the federal effort: "effective K-12 teacher education, engagement, undergraduate STEM education, and serving groups traditionally underrepresented in STEM fields."[70] The status report notes that strong arguments can be made for other STEM education policy areas, but states that these four were chosen as the priority areas for enterprise-wide coordination (agencies may still maintain their own STEM education priorities as well) because they represent the convergence of "national needs, Presidential priorities, and federal assets."[71]

To further enhance coordination at the federal level, some advocates maintain that Congress consider creating an Office of STEM Education and designating an Assistant Secretary for STEM Education at ED. Advocates for this approach claim that it would raise the profile of STEM education and improve administration of the various programs and policies at ED.[72]

Elementary and Secondary Schooling

Policymakers often express interest in making improvements in the early part of the STEM education pipeline—e.g., from kindergarten to 12th grade (also referred to as the "K-12" pipeline). Some analysts assert that mathematics and science achievement will not easily be raised unless the quality of K-12 teaching is improved. Other observers suggest low or unequal student achievement may be best addressed by adjustments to the K-12 system's accountability structure and standards for performance.

Teacher Quality

To many observers, mathematics and science teachers' lower likelihood of possessing subject-specific professional credentials, compared to teachers of other subjects, identifies a deficit of mathematics and science teacher quality. Although most teaching positions may be staffed, the K12 system's stock of fully credentialed mathematics and science teachers is in short supply.[73] A variety of solutions to the shortage of STEM teachers have been proposed.[74]

One set of proposals is directed at teachers currently in the classroom, while another set of solutions targets new or prospective teachers.

Some advocates feel that it is important to focus on performance, and that current teachers who are less than fully effective in the classroom are not provided the support and training they need to succeed and want to see federal funding for professional development (PD) specifically designed for STEM teachers maintained and expanded.[75] Other stakeholders propose establishing a STEM Master Teacher Corps that would reward experienced and effective mathematics and science teachers with increased career prestige and pay in return for mentoring and providing PD for less effective teachers.[76] Still others support increased use of online education, especially in rural schools that struggle to attract new teachers in any subject.[77] Additionally, some researchers support reforms that would ease the removal of ineffective teachers who do not sufficiently improve with PD and may not be cut out for the profession.[78]

Other stakeholders think improved recruitment and retention of high-quality new teachers is the primary solution to the mathematics and science teacher quality problem. Many who take this approach argue that federal teacher policies should assist state and local efforts to improve nontraditional routes to teaching—e.g., alternative certification policies and incentives for mid-career transition of STEM professionals.[79] Analysts have identified options for attracting new STEM teachers through traditional preparation programs by subsidizing their education costs—e.g., through direct grants, student loan repayment, or tax credits[80]—and encouraging colleges and universities to develop concurrent STEM and teaching degree programs.[81] Such recruitment strategies may also serve as retention tools when paired with requirements that new teachers fulfill service agreements. Some proponents prefer policies designed to attract and retain STEM teachers through financial incentives such as differential pay, housing subsidies, and signing bonuses.

Alternatively, some education analysts have criticized the federal policy focus on teacher quality, as measured by credentials, calling into question its link to student achievement and advocating instead for proposals to improve teacher effectiveness.[82] Specifically, those in this camp suggest reforming teacher evaluation systems to identify multiple levels of effectiveness; rewarding those at the top with performance pay and removing those at the lowest level of performance.[83] Related proposals would reform the structure of teacher preparation by rewarding teacher training programs, both traditional and non-traditional, on the basis of their graduates' classroom effectiveness rather than on certification exam results.[84]

Accountability and Standards

The accountability movement has been a powerful force in the federal education policy debate for some time, reaching particular prominence with the passage of the No Child Left Behind Act of 2001 (NCLB, P.L. 107-110).[85] NCLB reforms were based on the premise that, to improve outcomes from the K-12 system, student achievement must be accurately assessed and schools must be held accountable for measurable results. The law required that states establish achievement benchmarks, set annual goals (referred to as Adequate Yearly Progress, or AYP), and have all students reach "proficiency" in reading and mathematics by 2014.[86] NCLB also requires that students be assessed for their academic proficiency in science. However, these results are not tied to the accountability system. Some stakeholders are in favor of amending the law to mandate that schools ensure students also be proficient in science.[87]

Independent of federal involvement, states have begun development of a so-called common core of academic standards across the K-12 system.[88] This effort intends to create nationally consistent standards of knowledge and skills that students need in order to graduate from high school and succeed in entry-level, credit-bearing college courses or workforce training programs. So far, standards have been developed for reading and mathematics and adopted by 45 states and the District of Columbia.[89] Pointing to the perceived "mediocre" state of current state science standards, some analysts say the inclusion of science in the common core "is long overdue."[90]

Other K-12 Policy Issues

Some analysts argue that the current "STEM for all" approach is not working.[91] Those in this camp urge policymakers to focus limited federal resources on high-achieving students with an interest in STEM by, among other things, using federal education funding to create new specialty STEM high schools.[92] Other analysts seek to expand programs such as Advanced Placement and International Baccalaureate (AP/IB)—including grants to pay the AP/IB testing fees of low-income students—or seek to increase STEM education achievement among demographic groups with historically low participation rates in STEM fields.[93] (See section on "Broadening Participation.")

Post-Secondary Education

As a proportion of all federal STEM education funding, the majority of the federal investment in STEM education supports undergraduate, graduate, and post-graduate education and research. In each of the recent program inventories, post-secondary education accounted for more than half of the federal STEM education portfolio. However, some analysts argue that current U.S. STEM degree production rates are insufficient.[94] Proposals to improve post-secondary STEM education include those that seek to address remediation in the early college years or increase retention rates in STEM majors through graduation. Other proposals seek increased support for graduate study and post-doctoral research. Some analysts favor lowering barriers for foreign STEM students seeking entry into U.S. institutions of higher education.

Remediation and Retention

Researchers cite poor pre-college mathematics and science preparation and high rates of attrition among STEM majors as two major challenges for undergraduate STEM education in the United States.[95] In addition to the K-12 improvements discussed above, some observers propose additional federal investment in remedial education for students as they enter college. For example, some stakeholders advocate for increased funding for ED's Upward Bound Mathematics and Science program.[96] Others analysts see community colleges playing an important role in counteracting the perceived failings of secondary schools. For example, some stakeholders have called for partnerships between business and two-year colleges to prepare students for STEM careers.[97] Other analysts argue that proprietary, non-degree-granting institutions are well suited to provide STEM remediation and training.[98]

Some policymakers are concerned with low retention rates at undergraduate STEM programs. Although attrition in STEM fields may be due, in part, to poor K-12 preparation and to overall college attrition patterns, there are likely multiple reasons why students complete a non-STEM degree after showing initial interest in STEM.[99] Some analysts advocate for STEM education research programs that focus on improving undergraduate STEM education practices for all students, such as NSF's Widening Implementation and Demonstration of Evidence-based Reforms (WIDER) and Transforming Undergraduate Education in Science, Technology, Engineering, and Mathematics (TUES) programs.[100] Others support efforts to improve retention among groups traditionally underrepresented in STEM fields (including ethnic and racial minorities, the disabled, and females).[101]

Broadening Participation of Underrepresented Populations

The demographic profile of the U.S. student-age population is changing. The youth population is more racially and ethnically diverse than previous generations of Americans. In addition, more than half of U.S. college students are now female, and over half of all bachelor's, master's, and doctoral degrees awarded in the United States go to women.[102] Some observers say that these trends are problematic for the U.S. scientific and technological enterprise, which has historically relied mostly on a white male labor supply (particularly in fields such as mathematics and engineering). However, because the growth in the student-age population (and therefore future labor supply) is in segments that have typically been underrepresented in STEM,[103] these observers argue that underrepresented groups "embody a vastly underused resource and a lost opportunity for meeting our nation's technology needs."[104] The solution to this challenge, many stakeholders argue, is to increase (or broaden) the participation of women and ethnic and racial minorities in STEM education and employment.[105]

General agreement about the problems posed by racial, ethnic, and gender disparities in STEM education and employment has not translated into widespread agreement on either the causes of underrepresentation or policy solutions. Further, causes and solutions may be different for different population subsets.

Race and Ethnicity

Researchers have identified dozens of school and non-school variables that may contribute to racial and ethnic achievement gaps in STEM. For example, in 2011 researchers reviewed over 400 books, book chapters, journal articles, and policy reports on factors that influence minority student success in STEM (hereinafter referred to as the "2011 review").[106] The 2011 review found that the following factors positively influence the success of minority students in STEM:

- *K-12*—parental involvement and support, availability of bilingual education, culturally relevant pedagogy, early exposure to STEM fields, interest in STEM careers, self-efficacy in STEM subjects,[107] and STEM-related educational opportunities and support programs.[108]

The 2011 review also identified the following school-based factors as contributing to minority under-preparedness in elementary and secondary STEM education:

- *K-12*—a lack of resources (underfunding) and less qualified teachers at schools that serve minority students, limited access to Advanced Placement courses, disproportionate tracking of minority students into remedial education, teachers' low expectations, stereotype threat,[109] racial oppression and oppositional culture,[110] and premature departure from high school.[111]

At the post-secondary education level, the 2011 review identified the following factors as associated with varying levels of minority student STEM success in college:

- *Higher Education*—colorblind meritocracy, financial aid and employment, institutional type, campus culture and climate, institutional agents, self-concept and self-efficacy, and STEM opportunity and support programs.[112]

In addition to these school-based factors described in the 2011 review, other researchers have identified non-school factors that contribute to achievement gaps in both STEM and non-STEM fields. Some scholars argue that these non-school factors have been overlooked and that too much emphasis is placed on schools.[113] Non-school factors that have been identified as contributing to achievement gaps include concentrated poverty and single-parent households,[114] early childhood development,[115] and health.[116]

Policy solutions for broadening participation in STEM are also numerous. In 2010, Congress directed the National Academies to examine diversity in the STEM workforce and make recommendations for broadening participation.[117] Of the many recommendations in the resulting report, the Academies identified several policy options as top priorities. These include increased financial support for minority undergraduate STEM students, improved teacher preparation, more and better advanced courses and academic advising for minority K-12 students, improved transition to graduate school for minority undergraduates in STEM fields, and increased availability of research assistantships for minority graduates students in STEM.[118] Other researchers have proposed solutions such as charter schools and school choice,[119] faith-based schooling,[120] improved transfer pathways from community colleges and reducing undergraduate debt,[121] and participation in undergraduate research.[122] Some analysts propose increased investments in minority serving institutions (MSIs)—such as Historically Black Colleges and Universities (HBCU) or Tribal Colleges and Universities (TCUs)—as an option to broaden participation in STEM fields.[123]

Gender

Although the number of women earning colleges degrees has been increasing, they hold less than a quarter of STEM jobs nationally.[124] Scholars debate the causes of gender disparities in STEM. Some analysts assert that self-efficacy, institutional culture, discrimination, and bias limit female participation in science.[125] Other observers do not find evidence of widespread, contemporary discrimination against women in STEM fields; instead, they primarily attribute disparities to family formation and child rearing, gendered expectations, lifestyle choices, career preferences, and personal choice, among other complex factors.[126]

Differences in beliefs about the causes of gender disparities in STEM lead to different emphases in proposed solutions. Scholars who generally align with the discrimination hypothesis suggest a variety of policy options. Among these are policies that seek to increase girls' interest in STEM; create college environments (e.g., institutional culture) that attract and retain female students and faculty; and counteract bias by, among other things, creating clear and transparent criteria for success.[127] Scholars who generally align with the preferences hypothesis recommend so-called family friendly policies at academic institutions (e.g., part-time tenure track positions and childcare) and propose federal funding for research "on the differing lifecourses of women's and men's careers to determine whether the traditional timing of hiring, tenure and promotion may deny society and science the contributions of talented women."[128]

Other Factors

Some researchers argue that income is the most critical variable in achievements gaps and that gaps between children from high- and low-income families have grown substantially in recent decades. The income achievement gap, these researchers argue, is as determinative (if not more) than race.[129] Researchers have identified summer learning loss as one of the possible contributors to achievement gaps by income.[130] Studies show that students lose skills over the summer, especially in mathematics, and that the effects of these losses appear to accumulate over time. Further, losses appear to disproportionately affect low-income students.[131]

Other researchers have observed mathematics achievement gaps by urbanization level (e.g., between rural, urban, and suburban youth), finding that urban and rural youth have lower average mathematics achievement levels than their suburban peers and that this gap appears to widen between kindergarten and 8th grade. These findings, say the researchers, are not solely attributable to differences in socio-economic status.[132]

APPENDIX A. DATA SOURCES AND MAJOR PUBLICATIONS

Federal STEM education analysts rely on a number of sources and major publications for data about the federal STEM education effort and the condition of STEM education in the United States and around the globe. This appendix includes links to sources and publications where readers can find the most up-to-date STEM education data and information.

National and International Assessments

- **National Assessment of Educational Progress (NAEP)**—The NAEP is the largest nationally representative and continuing assessment of U.S. K-12 students. There are two NAEP assessments: Main NAEP and Long-Term Trends (LTT). The Main NAEP administers assessments in 12 subject areas, including mathematics and science. The LTT assesses mathematics and reading. More information about these assessments is available at
    - *Main NAEP Science*—http://nces.ed.gov/nationsreportcard/science/.
    - *Main NAEP Mathematics*—http://nces.ed.gov/nationsreportcard/mathematics/.
    - *LTT Mathematics*—http://nationsreportcard.gov/ltt_2008/.
- **Trends in International Mathematics and Science Study (TIMSS)**—TIMSS is an international test that assesses the mathematics and science achievements of U.S. 4th and 8th grade

students in a manner that is comparable across countries. More information about TIMSS is available at http://nces.ed.gov/timss/.

- **Program for International Student Assessment (PISA)**—PISA assesses the reading, mathematics, and science literacy of 15-year-old students in dozens of industrialized and developing nations. More information about PISA is available at http://nces.ed.gov/surveys/pisa/.

Federal Programs and Inventories

- U.S. Government Accountability Office, Science, Technology, Engineering, and Mathematics Education: Strategic Planning Needed to Better Manage Overlapping Programs Across Multiple Agencies (GAO-12-108)—This 2012 GAO report on federal STEM education programs includes an inventory of federal STEM education programs and policy recommendations. The report is available at http://gao.gov/products/GAO-12-108.
- National Science and Technology Council, A Report from the Federal Inventory of STEM Education Fast-Track Action Committee (NSTC 2011) and A Report from the Federal Coordination in STEM Education Task Force (NSTC 2012)—These reports provide an inventory of federal STEM education investments from FY2008 to FY2010 actual (in the 2011 report) and an update with information from FY2011 enacted to FY2013 requested (in the 2012 report).
    - NSTC 2011—This report is available at http://www.whitehouse.gov/sites/default/files/microsites/ostp/costem__federal_stem_education_portfolio_report.pdf.
    - NSTC 2012—This report is available at http://www.whitehouse.gov/sites/default/files/microsites/ostp/nstc_federal_stem_education_coordination_report.pdf.
- U.S. Department of Education, Academic Competitiveness Council (ACC), Report of the Academic Competitiveness Council—The 2007 ACC report provides an inventory of federal STEM education programs with funding data from FY2005 actual to the FY2007 President's budget request. Includes policy recommendations and an assessment of STEM education program evaluations from across the federal enterprise. The 2007 ACC report is available at

http://www2.ed.gov/about/inits/ed/competitiveness/acc-mathscience/index.html.

- U.S. Government Accountability Office, Higher Education: Federal Science, Technology, Engineering, and Mathematics Programs and Related Trends (GAO-06-114)—This 2005 GAO report includes an inventory of federal STEM education programs and assesses program goals and constituencies served. The 2005 GAO report is available at http://www.gao.gov/products/GAO-06-114.

Condition of STEM Education

- **National Science Board, *Science and Engineering Indicators 2012***—This publication provides, among other things, one of the most comprehensive collections of key STEM indicators. It is published every two years. More information about *Science and Engineering Indicators* is available at http://www.nsf.gov/statistics/seind12/front/fronts6.htm.
- **National Science Foundation, National Center for Science and Engineering Statistics (NCSES)**—NCSES compiles and analyzes a variety of STEM data. Much of this data may be found in *Science and Engineering Indicators*, but the NCSES website includes separate, detailed, and timely publications on various STEM education data. More information about NCSES is available at http://www.nsf.gov/statistics/.

Appendix B. Selected Major Legislation

Depending on how broadly the term is defined, federal interest in STEM education may be traced to the 1st Congress. Several institutions that would become central parts of the federal STEM education effort—such as Health and Human Services (1798, 1 Stat. 605),[133] the Smithsonian Institution (1846, 9 Stat. 103), the National Academy of Sciences (1863, 12 Stat. 806), and Department of Education (originally the Office of Education, 1867, 14 Stat. 434)—were in place before the United States celebrated its first centennial.

Federal STEM education policymaking intensified after World War II. The desire to maintain the scientific achievements of the war led to the creation of the National Science Foundation in 1950. By 1952, NSF was issuing GRF awards to promising STEM graduate students. The Soviet Union's launch of the first artificial satellite, *Sputnik*, triggered fears that the United States was falling behind in mathematics and science education and led to the National Defense Education Act of 1958, which some analysts cite as the first federal foray into STEM education policy in the modern era.

This appendix includes selected historical federal STEM education measures arranged by date.[134]

Land Ordinance of 1785[135] and Northwest Ordinance of 1787[136]

The Land Ordinance of 1785 was one of a series of three measures providing for the political and geographic incorporation of the Northwest Territories in the Union. These measures were passed by the Continental Congress after the Revolutionary War and prior to the adoption of the Constitution. Drafted primarily for the purpose of disposing of land in the territories, the Land Ordinance of 1785 directed surveyors to establish townships in the territories. These townships were to be subdivided into lots, one of which (lot number 16) was to be preserved for the maintenance of a public school. The Land Ordinance's more famous cousin, the Northwest Ordinance of 1787, established governments in the territories and provided for the civil liberties of the inhabitants. On the question of education the Northwest Ordinance said, "Religion, morality and knowledge, being necessary to good government and the happiness of mankind, schools and the means of education shall forever be encouraged."[137]

Marine Hospital Service Act of 1798 (1 Stat. 605)[138]

Congress established the Marine Hospital Service (MHS) in 1798 to provide medical care for merchant seaman. Many federal health agencies trace their origin to the establishment of the MHS; including the National Institute of Health (NIH), which began as the Hygienic Laboratory within the MHS in 1887. The Ransdell Act of 1930 (P.L. 71-251) re-designated the Hygienic Laboratory as the NIH and authorized fellowships at the institute. Although NIH education and training funding in the 1930s, 1940s, and 1950s primarily focused on post-doctoral researchers and clinical traineeships for physicians, in 1948 the National Institute of Cancer began awarding funds to institutions to improve undergraduate education.[139] In 1974 Congress established the

National Research Service Award (NRSA) at NIH. The National Research Service Award Act (P.L. 93-348) consolidated and established under a single authority existing research and fellowship authorities. P.L. 107-206 re-named the NRSA the "Ruth L. Kirschstein National Research Service Award" in 2002.

Morrill Acts of 1862 (12 Stat. 503)[140] and 1890 (26 Stat. 417)

The Morrill Act of 1862 authorized the sale of federal lands and distribution of the proceeds to the states for the purpose of establishing colleges in the "mechanic arts" (e.g., engineering, manufacturing, inventions), agriculture, and military tactics. The original Morrill Act did not apply to the "states in rebellion," but in 1890 Congress passed a subsequent measure to provide for colleges in Southern states.[141] The 1890 Morrill Act also expanded the purposes of the colleges to include "agriculture, the mechanic arts, the English language, and the various branches of mathematical, physical, natural, and economic science." These provisions were repealed in 1981 and replaced with "food and agricultural sciences."[142] Colleges funded by these acts include the Massachusetts Institute of Technology, Clemson University, and many U.S. Historically Black Colleges and Universities.[143]

National Science Foundation Authorization Act of 1950 (P.L. 81-507)[144]

The NSF was established in 1950 to—in part—"develop and encourage the pursuit of a national policy for basic research and education in the sciences."[145] Congress passed the act authorizing the Foundation after several years of debate and a veto in 1947.[146] NSF distributed its first fellowships to pre- and post-doctoral STEM students in 1952. As early as 1953 the Foundation began supporting teacher institutes as a means of improving STEM education in the lower grades.[147] Although both the Congress and the President have made changes to the NSF since its founding, STEM education has remained a core function of the agency.

National Defense Education Act of 1958 (P.L. 85-864)

Passed in 1958 in response to the Soviet Union's launch of *Sputnik*, the National Defense Education Act (NDEA), sought to address concerns about "existing imbalances in our educational programs which have led to an insufficient proportion of our population educated in science, mathematics, and modern foreign languages and trained in technology."[148] Among its many provisions, the NDEA authorized the first federal student loan program; provided funds to states for science, mathematics, and modern foreign language instruction; and authorized grants to states for programs to identify and encourage gifted students. Some NDEA scholars assert that this act paved the way for the Elementary and Secondary Education Act of 1965 by establishing a legislative precedent for federal education aid to states.[149]

Elementary and Secondary Education Act of 1965 (P.L. 89-10)[150]

The primary source of federal aid to K-12 education is the Elementary and Secondary Education Act (ESEA).[151] ESEA was initially enacted in 1965 and was most recently amended and reauthorized by the No Child Left Behind Act of 2001 (NCLB, P.L. 107-110). STEM education was not central to the ESEA as originally constructed in 1965, but STEM-specific provisions have been added in subsequent reauthorizations. For example, as amended by No Child Left Behind, the act authorizes the Mathematics and Science Partnerships (MSP) program at ED[152] and requires states to have mathematics assessments and standards.

Higher Education Act of 1965 (P.L. 89-329)

The Higher Education Act (HEA) authorizes a series of programs that provide federal aid and support to institutions of higher education as well as a broad array of federal student aid programs that assist students and their families with paying for or financing the costs of obtaining a postsecondary education. The HEA was most recently amended in 2008 by the Higher Education Opportunity Act (P.L. 110-315); however, the only major STEM-related postsecondary program administered by ED was enacted by the Higher Education Reconciliation Act of 2005 (HERA; P.L. 109-171). The HERA amendments included authorization and appropriations for the SMART Grant program, which provided $4,000 grants to students majoring in STEM fields. Congress provided that the program sunset at the end of the 2010-2011 academic year. Approximately $1.4 billion in grants were awarded between FY2006 and FY2010.

Department of Education Organization Act of 1979 (P.L. 96-88)

The Department of Education Organization Act established ED as an independent federal agency. Section 304 of the act transferred science education programs established at NSF to ED.[153] Excluded from this directive were programs that related to scientific career development, continuing education of scientific personnel, career-focused broadening participation programs, research and development in science learning, and programs to inform the general public about the nature of science and technology and related policy issues. The conference report on the final bill included two specific examples of NSF programs to be transferred to ED: Elementary and Pre-school Science Teacher Training and Minority Institutions Science Improvement.[154] The act provided only for the transfer of programs in existence at the time of enactment and included a provision affirming NSF's authority to initiate and conduct programs under its originating act (P.L. 81-507).

Education for Economic Security Act of 1984 (P.L. 98-377)

The Education for Economic Security Act of 1984 (EESA) authorized teacher institutes and mathematics and science education development programs (including partnerships) at the National Science Foundation; directed the Department of Education to provide grants to states and local educational agencies for STEM teacher training and development; and authorized presidential awards for teaching in mathematics and science, among other things. EESA was enacted following publication of several reports—most notably *A Nation at Risk*—that were highly critical of the U.S. education system and amid growing concerns about international competitiveness in the wake of the 1970s recession and apparent ascendancy of the Japanese and German economies.

America COMPETES Act of 2007 (P.L. 110-69) and America COMPETES Reauthorization Act of 2010 (P.L. 111-358)[155]

The America COMPETES Act (and its 2010 reauthorization) authorized a variety of STEM education programs at several federal science agencies and ED. Most of the specified STEM education appropriations authorizations in these acts are at the NSF, but the acts also contain STEM education provisions for the Department of Energy (DOE), the National Aeronautics and Space Administration (NASA), and the National Oceanic and Atmospheric Administration (NOAA). Among other things, the 2010 reauthorization established a federal government-wide STEM education coordinating committee under the National Science and Technology Council.

End Notes

[1] Earlier examples include debate at the Constitutional Convention about whether to empower the federal government "to establish seminaries for the promotion of literature and the arts and sciences." James Madison, "Saturday, August 18," *Notes of Debates in the Federal Convention of 1787*, TeachingAmericanHistory.org website.

[2] U.S. President George Washington, *First Annual Message to Congress on the State of the Union, January 8, 1790*, The American Presidency Project website.

[3] The term "scientific thinking" has many definitions. In general, it refers to the skills, processes, and methods of science (broadly defined).

[4] Although a global competitiveness rationale drives much of the contemporary debate about STEM education policies, STEM illiteracy (particularly innumeracy) has also been linked to other national challenges such as the mortgage crisis and even medication errors. For example, see Kristopher Gerardi et al., *Financial Literary and Subprime Mortgage Delinquency: Evidence from a Survey Matched to Administrative Data*, Working Paper 2010-10, Federal Reserve Bank of Atlanta, April 2010; and Robert Preidt, "Parents' Poor Math Skills May = Medication Errors," National Institutes of Health, U.S. National Library of Medicine, Medline Plus website, April 30, 2012.

[5] Another source of STEM labor in the United States is immigration. For more information about foreign STEM workers, see CRS Report R42530, *Immigration of Foreign Nationals with Science, Technology, Engineering, and Mathematics (STEM) Degrees*, by Ruth Ellen Wasem.

[6] This is a rough estimate. The limitations of this calculation are explained in the section on "The Federal Effort in STEM Education."

[7] The America COMPETES Act of 2010 defines the term STEM for the agencies it authorizes, including the NSF. As defined by P.L. 111-358, Section 2, the term STEM means "the academic and professional disciplines of science, technology, engineering, and mathematics." In practice, NSF funds research in the so-called core sciences (e.g., mathematics and physical sciences) and engineering as well as psychology and the social sciences.

[8] While the DHS definition of a STEM field is, in general, more narrow than that of the NSF, DHS announced in May 2012 that it was expanding the list of fields it would support to include pharmaceutical sciences, econometrics, quantitative economics, and others. U.S. Department of Homeland Security, "DHS Announces Expanded List of STEM Degrees," press release, May 11, 2012. See also, U.S. Department of Homeland Security, Immigration and Customs and Enforcement, *STEM-Designated Degree Program List: 2012 Revised List*, 2012.

[9] Jean Moon and Susan Rundell Singer, "Bringing STEM into Focus," *Education Week*, vol. 31, no. 19 (February 1, 2012), pp. 32, 24.

[10] National Academy of Sciences, National Academy of Engineering, and Institute of Medicine, Committee on Prospering in the Global Economy of the 21st Century: An Agenda for American Science and Technology, and Committee on Science, Engineering, and Public Policy, *Rising Above the Gathering Storm: Energizing and Employing America for a Brighter Economic Future* (Washington, DC: National Academies Press, 2007).

[11] The ACC was created by the Deficit Reduction Act of 2005 (P.L. 109-171) and charged with conducting a yearlong study to identify all federal STEM education programs. U.S. Department of Education, *Report of the Academic Competitiveness Council*, May 2007.

[12] President Bill Clinton established the NSTC by Executive Order 12881 on November 23, 1993. The NSTC aims to coordinate science and technology policy across the federal government. For more information on the NSTC, see CRS Report RL34736, *The President's Office of Science and Technology Policy (OSTP): Issues for Congress*, by John F. Sargent Jr. and Dana A. Shea.

[13] U.S. Government Accountability Office, *Federal Science, Technology, Engineering, and Mathematics Programs and Related Trends*, GAO-06-114, Washington, DC, October 2005.

[14] In 2010, using a method similar to that of the ACC, the Office of Management and Budget (OMB) found 171 federal STEM education programs funded at about $3.8 billion. Unpublished data from the OMB. Available upon request.

[15] Executive Office of the President, National Science and Technology Council, Committee on STEM Education, Federal Inventory of STEM Education Fast-track Action Committee, *The Federal Science, Technology, Engineering, and Mathematics (STEM) Education Portfolio*, December 2011.

[16] U.S. Government Accountability Office, *Science, Technology, Engineering, and Mathematics Education: Strategic Planning Needed to Better Manage Overlapping Programs across Multiple Agencies*, GAO-12-108, January 2012.

[17] As enacted by the Higher Education Reconciliation Act of 2005 (P.L. 109-171), the SMART Grant program awarded $4,000 grants to students majoring in STEM fields. Congress provided that the program sunset at the end of the 2010- 2011 academic year. Approximately $1.4 billion in grants were awarded between FY2006 and FY2010.

[18] NSTC-2011, p. 16.

[19] GAO-2005, p. 13.

[20] This amount includes FY2012 enacted funding for Institutional Research Training Grants and individual fellowships. Executive Office of the President, National Science and Technology Council, Committee on STEM Education, Federal Coordination in STEM Education Task Force, *Coordinating Federal Science, Technology, Engineering, and Mathematics (STEM) Education Investments: Progress Report*, February 2012, p. 36.

[21] FY2012 funding for the NRSA is $273.5 million. Total STEM education funding at NIH, according to the NSTC2011 inventory (updated), is $560.4 million. See Executive Office of the President, National Science and Technology Council, Committee on STEM education,

Federal Coordination in STEM Education Task Force, *Coordinating Federal Science, Technology, Engineering, and Mathematics (STEM) Education Investments: Progress Report*, February 2012, p. 36.

[22] NRSA offers several types of Institutional Research Training Grants. Institutional eligibility varies.

[23] More information about the NRSA is available at http://grants.nih.gov/training/nrsa.htm.

[24] National Science Foundation, *FY2013 Budget Request to Congress*, February 13, 2012, pp. NSF-wide Investments-68-69. GRF program rules require institutions of higher education to exempt GRF fellows from tuition and fees. The COE provides funds to the institution for the cost of educating the student. The institution is responsible for tuition and fees in excess of the COE. Stipends are a form of salary and may be treated as taxable income.

[25] Ibid. NSF raised the COE from $10,500 to $12,000 in FY2012. For FY2013, the NSF seeks to increase the stipend level to $32,000. The FY2013 NSF budget request also seeks to increase the number of awards to 8,900 (2,000 new).

[26] Between FY2003 and FY2008, funds for the GRF came principally from NSF's main education account. Funding levels for the GRF during this period ranged between $85 million and $96 million. Starting in FY2009, the NSF increased the main research account contribution to the GRF program from less than $10 million per year to between $34 and $56 million per year. For FY2013 the NSF seeks a total of $243 million for the GRF. The main research and education accounts would each provide about half of this amount.

[27] Robert Atkinson et al., *Innovation Policy on a Budget: Driving Innovation in a Time of Fiscal Constraint*, Information Technology and Innovation Foundation, September 2010, p.10.

[28] In its report on legislation authorizing the MSP program at NSF, the House Committee on Science noted "The Committee believes that the Partnership program in this Act is complementary to, and not duplicative of a similarly titled math and science partnership program in H.R. 1, 'The No Child Left Behind Act of 2001.' ... The Committee anticipates that the two programs will draw on each others' strengths and that the most promising NSF-funded projects will be used as models and brought to full scale by the Department of Education's partnership program." See H.Rept. 107-134.

[29] For more information on NSF's MSP program, see National Science Foundation, Directorate for Education and Human Resources, *National Impact Report: Mathematics and Science Partnership Program*, NSF 2010-046, 2010, p. 18.

[30] Julie Siebens, *Science and Engineering Degrees: 2009*, U.S. Department of Commerce, Economics and Statistics Administration, U.S. Census Bureau, American Community Survey Briefs, ACSCR/09-14, October 2010.

[31] The low was 30.5% in 1991 and the high was 35.6% in 1968 and 1970. National Science Foundation, National Center for Science and Engineering Statistics, "Table 1," *S&E Degrees: 1966-2008*, NSF 11-316, June 2011.

[32] Ibid. The low was 56.1% in 1976 and the high was 67.5% in 2008.

[33] Ibid.

[34] Kelly Kang, NCSES *Infobrief: Graduate Enrollment in Science and Engineering Grew Substantially in the Past Decade but Slowed in 2010*, National Science Foundation, National Center for Science and Engineering Statistics, NSF 12-317, May 2012.

[35] Out of a possible 500 points, the average scores of 4th graders have gone from 213 in 1990 to 241 in 2011; 8th graders have gone from 263 to 284. The 1990 scores are statistically significant ($p < .05$) from 2011. U.S. Department of Education, Institute of Education Sciences, National Center for Education Statistics, *The Nation's Report Card: Mathematics 2011*, NCES 2012-458, November 2011; and, U.S. Department of Education, Institute of

Education Sciences, National Center on Education Statistics, *2011 Condition of Education*, NCES 2011-033, May 2011, p. 46.

[36] Comparable data for the NAEP science assessment are not available. The science assessment was changed in 2009 to reflect changes in curriculum standards, assessments, research, and science. As such, the 2009 results are not comparable with results from prior years.

[37] The 2011 gap between the average scores of Hispanics and whites on the NAEP mathematics test was 20 points; for black and white students, the gap was 25 points. For more information on NAEP results and scoring, see U.S. Department of Education, Institute of Education Sciences, National Center for Education Statistics, *The Nation's Report Card: Mathematics 2011*, NCES 2012-458, November 2011, p. 13.

[38] National Science Foundation, National Center for Science and Engineering Statistics, "Table 11," *S&E Degrees: 1966-2008*, NSF 11-316, June 2011, p. 15.

[39] The rates for Hispanic/Latino, American Indian/Alaska Native, and black/African American S&E graduate enrollments between 2000 and 2010 were 65%, 55%, and 50%, respectively. Kelly Kang, NCSES *Infobrief: Graduate Enrollment in Science and Engineering Grew Substantially in the Past Decade but Slowed in 2010*, National Science Foundation, National Center for Science and Engineering Statistics, NSF 12-317, May 2012.

[40] National Science Foundation, National Center for Science and Engineering Statistics, "Table 11," *S&E Degrees: 1966-2008*, NSF 11-316, June 2011, p. 15.

[41] For more information on student achievement, see CRS Report R40514, *Assessment in Elementary and Secondary Education: A Primer*, by Rebecca R. Skinner.

[42] Michael B. Allen, *Eight Questions on Teacher Preparation: What Does the Research Say?*, Education Commission of the States, July 2003.

[43] U.S. Department of Education, Institute of Education Sciences, National Center for Education Statistics, "Table 2," *Education and Certification Qualifications of Departmentalized Public High School-Level Teachers of Core Subjects: Evidence from the 2007-08 Schools and Staffing Survey*, NCES 2011-317, May 2011, p. 14.

[44] Organization for Economic Cooperation and Development, *PISA 2009 Results: What Students Should Know and Can Do: Volume I*, December 2010.

[45] Ina V.S. Mullis et al., *TIMSS 2007 International Mathematics Report: Findings from IEA's Trends in International Mathematics and Science Study at the Fourth and Eighth Grades*, Boston College, Lynch School of Education, TIMSS & PIRLS International Study Center, 2008; and Michael O. Martin, et al., *TIMSS 2007 International Science Report: Findings from IEA's Trends in International Mathematics and Science Study at the Fourth and Eighth Grades*, Boston College, Lynch School of Education, TIMSS & PIRLS International Study Center, 2008.

[46] Some research has found that U.S. students do not try very hard on low-stakes standardized tests and that this affects scores. For example, research on financial incentives to improve student performance found that "The large effects of these relatively modest financial incentives [$10 to $20] suggest that at baseline this population of students [e.g., students in the study sample] puts forth low effort in response to low (perceived) returns to achievement on standardized tests." Steven D. Levitt et. al., *The Behavioralist Goes to School: Leveraging Behavioral Economics to Improve Educational Performance*, National Bureau of Economic Research, Working Paper no. 18165, June 2012.

[47] National Association of Secondary School Principals, "NASSP Responds to International Assessment Results," press release, December 10, 2010; and Cynthia McCabe, "The Economics Behind International Education Rankings," *NEAToday*, December 9, 2010.

[48] For more information on issues related to foreign students and foreign technical workers, see CRS Report R42530, *Immigration of Foreign Nationals with Science, Technology, Engineering, and Mathematics (STEM) Degrees*, by Ruth Ellen Wasem; and CRS Report 97-746, *Foreign Science and Engineering Presence in U.S. Institutions and the Labor Force*, by Christine M. Matthews.

[49] In 2009, there were 611,629 graduate students in science and engineering fields in the United States. Of these, 168,850 (27.6%) were temporary residents. National Science Board, *Science and Engineering Indicators: 2012*, NSB 12-01, National Science Foundation, January 13, 2012, p. 2-28.

[50] National Science Foundation, "Figure 3.7—Citizenship Status of Ph.D.'s: 1960-1999," *U.S. Doctorates in the 20th Century*, NSF 06-319, October 2006; and Mark K. Feigener, *Number of Doctorates Awarded in the United States Declined in 2010*, National Science Foundation, NSF 12-303, November 2011.

[51] The House Committee on the Judiciary examined foreign student policy issues in an October 5, 2011, hearing titled, "STEM the Tide: Should America Try to Prevent and Exodus of Foreign Graduates of U.S. Universities with Advanced Degrees?" A video of the hearing, as well as written testimony from witnesses, is available at http://judiciary.house.gov/hearings/hear_10052011_2.html.

[52] These totals include foreign students. China expanded its domestic production of S&E doctoral degrees from about 2,700 in 1994 to almost 28,500 in 2008. National Science Board, *Science and Engineering Indicators, 2012*, NSB 12- 01, National Science Foundation, January 13, 2012, pp. 2-5 and 2-34 to 2-37.

[53] Gary Gereffi et al., "Getting the Numbers Right: International Engineering Education in the United States, China, and India," *Journal of Engineering Education*, vol. 97, no. 1 (January 2008), p. 13-25.

[54] For more information on issues related to the U.S. STEM labor supply, see CRS Report RL34091, *Globalization, Worker Insecurity, and Policy Approaches*, by Raymond J. Ahearn; CRS Report RL32292, *Offshoring (or Offshore Outsourcing) and Job Loss Among U.S. Workers*, by Linda Levine; CRS Report R42141, *Computer-Related Occupations Under the Fair Labor Standards Act (FLSA)*, by Gerald Mayer; and CRS Report R42411, *The Tool and Die Industry: Contribution to U.S. Manufacturing and Federal Policy Considerations*, by Bill Canis.

[55] Multiple reports from a variety of respected U.S. academic, scientific, and business organizations have made this argument. For example, see National Academy of Sciences, National Academy of Engineering, and Institute of Medicine, Committee on Prospering in the Global Economy of the 21st Century: An Agenda for American Science and Technology, and Committee on Science, Engineering, and Public Policy, *Rising Above the Gathering Storm: Energizing and Employing America for a Brighter Economic Future* (Washington, DC: National Academies Press, 2007); and, Frederick M. Hess, Andrew P. Kelly, and Olivia Meeks, *The Case for Being Bold: A New Agenda for Business in Improving STEM Education*, U.S. Chamber of Commerce, Institute for a Competitive Workforce, April 2011.

[56] For example, see Richard Freeman, "The Market for Scientists and Engineers," *NBER Reporter*, no. 3 (Summer 2007), pp. 6-8; Ron Hira, "U.S. Policy and the STEM Workforce System," *American Behavioral Scientist*, vol. 53, no. 7 (March 2010), pp. 949-961; Testimony of Institute for the Study of International Migration Director of Policy Studies B. Lindsay Lowell in U.S. Congress, House Committee on the Judiciary, Subcommittee on Immigration Policy and Enforcement, *"STEM" the Tide: Should America Try to Prevent an*

Exodus of Foreign Graduates of U.S. Universities with Advanced Science Degrees?, hearings, 112th Cong., 1st sess., Serial No. 112-64, October 5, 2011; and B. Lindsay Lowell and Harold Salzman, *Into the Eye of the Storm: Assessing the Evidence on Science and Engineering Education, Quality, and Workforce Demand*, Urban Institute, October 2007.

[57] Anthony P. Carnevale, Nicole Smith, and Michelle Melton, *STEM: Science, Technology, Engineering, and Mathematics*, Georgetown University Center on Education and the Workforce, October 20, 2011, p.7.

[58] New York Times Editorial Board, "48th is Not a Good Place," *New York Times*, October 26, 2010, p. A28.

[59] For example, see Senator Tom Coburn, *The National Science Foundation: Under the Microscope*, April 2011, p. 54; and U.S. Department of Education, *A Blueprint for Reform: The Reauthorization of the Elementary and Secondary Education Act,* March 2010.

[60] This argument is, for example, part of the rationale for Administration-proposed program consolidations at ED. For more information, see CRS Report R41355, *Administration's Proposal to Reauthorize the Elementary and Secondary Education Act: Comparison to Current Law*, by Rebecca R. Skinner et al.

[61] This difficulty in driving a national agenda would depend on how the grants to states were structured. Federal policymakers could still drive a national STEM education agenda if they made receipt of consolidated program funds contingent on meeting certain defined national goals. However, some states may reject such efforts as overly prescriptive.

[62] For example, the National Aeronautics and Space Administration (NASA) has both unique workforce needs (e.g., astrobiologists) and unique assets that it can bring to the national STEM education effort (e.g., teaching from space).

[63] GAO states, "over 90% of STEM education programs that reported administrative costs estimated having administrative costs lower than 10% of their total program costs." U.S. Government Accountability Office, *Science, Technology, Engineering, and Mathematics Education: Strategic Planning Needed to Better Manage Overlapping Programs Across Multiple Agencies*, GAO-12-108, January 2012, p. 22.

[64] For example, programs that appear duplicative by some measures (e.g., target group) may have different intangible assets that could affect program implementation and outcomes. In this sense, they may not be *strictly* duplicative.

[65] U.S. Government Accountability Office, *Opportunities to Reduce Potential Duplication in Federal Teacher Quality Programs*, GAO-11-510T, April 13, 2011.

[66] Executive Office of the President, National Science and Technology Council, Committee on STEM Education, Fast-Track Action Committee on Federal Investments in STEM Education, *The Federal Science, Technology, Engineering, and Mathematics (STEM) Education Portfolio*, December 2011, p. 37.

[67] See U.S. Government Accountability Office, *Science, Technology, Engineering, and Mathematics Education: Strategic Planning Needed to Better Manage Overlapping Programs Across Multiple Agencies*, GAO-12-108, January 2012, p. 20.

[68] The NSTC-2011 inventory stated that "the primary issue [instead of duplication] is how to strategically focus the limited federal dollars available within the vast landscape of opportunity so they will have the most significant impacts possible in areas of national priority." (Executive Office of the President, National Science and Technology Council, Committee on STEM Education, Fast-Track Action Committee on Federal Investments in STEM Education, *The Federal Science, Technology, Engineering, and Mathematics (STEM) Education Portfolio*, December 2011, p. 37.) In its January 2012 report, GAO recommended not only that a federal STEM education strategy plan be drafted, but that

NSTC should also develop policies to ensure compliance. In particular, the GAO recommended that the NSTC develop (1) guidance for agencies on how to incorporate STEM education efforts into agency performance plans; (2) a framework for how agencies will be monitored to ensure they collect and report on strategic plan goals; and (3) guidance to help agencies determine the types of evaluations that may be feasible and appropriate for different types of STEM education programs. Additionally, GAO recommended that the NSTC work with agencies to identify programs that might be candidates for consolidation or elimination. (U.S. Government Accountability Office, *Science, Technology, Engineering, and Mathematics Education: Strategic Planning Needed to Better Manage Overlapping Programs Across Multiple Agencies*, GAO-12-108, January 2012, p. 31.)

[69] Executive Office of the President, National Science and Technology Council, Committee on STEM Education, Federal Coordination in STEM Education Task Force, *Coordinating Federal Science, Technology, Engineering, and Mathematics (STEM) Education Investments: Progress Report*, February 2012.

[70] Ibid. p. 13.

[71] Ibid. p. 17.

[72] Triangle Coalition for Science and Technology Education, "Doing What's Best for Science, Technology, Engineering, and Mathematics Education," talking points, January 2011.

[73] Richard M. Ingersoll and David Perda, *The Mathematics and Science Teacher Shortage: Fact and Myth*, Consortium for Policy Research in Education, CPRE Research Report #RR-62, Philadelphia, PA, March 2009.

[74] For a discussion of teacher issues, see CRS Report R41267, *Elementary and Secondary School Teachers: Policy Context, Federal Programs, and ESEA Reauthorization Issues*, by Jeffrey J. Kuenzi.

[75] For example, the Triangle Coalition for Science and Technology Education advocates for maintaining current funding for ED's MSP program and increased funding for professional development support under ED's Teacher Quality State Grant program. Triangle Coalition for Science and Technology Education, "Doing What's Best for Science, Technology, Engineering, and Mathematics Education," talking points, January 2011.

[76] Letter from STEM Education Coalition to Senators Tom Harkin and Michael B. Enzi, June 20, 2011.

[77] Jerry Johnson and Marty Strange, *Why Rural Matters 2009: State and Regional Challenges and Opportunities*, The Rural School and Community Trust, October 30, 2009.

[78] Saba Bireda, *Devil in the Details: An Analysis of State Teacher Dismissal Laws*, Center for American Progress, June 3, 2010.

[79] National Council on Teacher Quality, *Tackling the STEM Crisis*, Summer 2009.

[80] For more information about higher education grants, loans, and tax benefits, see CRS Report R42446, *Federal Pell Grant Program of the Higher Education Act: How the Program Works, Recent Legislative Changes, and Current Issues*, by Shannon M. Mahan; CRS Report R40122, *Federal Student Loans Made Under the Federal Family Education Loan Program and the William D. Ford Federal Direct Loan Program: Terms and Conditions for Borrowers*, by David P. Smole; and CRS Report R41967, *Higher Education Tax Benefits: Brief Overview and Budgetary Effects*, by Margot L. Crandall-Hollick.

[81] National Science Teachers Association, *NSTA Position Statement: Science Teacher Preparation*, July 2004.

[82] For example, see Dan Lips and Jena Baker McNeill, *A New Approach to Improving Science, Technology, Engineering, and Mathematics Education*, The Heritage Foundation, Backgrounder no. 2249, April 15, 2009.

[83] National Center on Performance Incentives, *Teacher Performance Pay: A Review*, November 2006.

[84] The Education Schools Project, *Educating School Teachers*, September 2006.

[85] More information on ESEA accountability can be found in CRS Report R41533, *Accountability Issues and Reauthorization of the Elementary and Secondary Education Act*, by Rebecca R. Skinner.

[86] More information assessment in ESEA can be found in CRS Report R40514, *Assessment in Elementary and Secondary Education: A Primer*, by Rebecca R. Skinner.

[87] NSTA Reports, "Should Science Count Toward AYP?," *National Science Teachers Association website—NSTA Web News Digest*, February 7, 2011.

[88] The Common Core State Standards Initiative is a state-led effort coordinated by the National Governors Association and the Council of Chief State School Officers. More information may be found at http://www.corestandards.org/.

[89] The National Research Council's Board on Science Education is currently developing a conceptual framework to guide the development of new science education standards. More information may be found at http://www7.nationalacademies.org/bose/Standards_Framework_FAQs.html.

[90] Lawrence S. Lerner et al., *The State of State Science Standards 2012*, Thomas B. Fordham Institute, January 31, 2012, p. 4-5.

[91] The "STEM for all" approach asserts that STEM competencies are central to contemporary work, life, and citizenship and that all U.S. students should have some mastery of these subjects and skills. In general, a "STEM for all" approach seeks to distribute STEM education resources widely, across all student skill levels, rather than targeting federal resources at high-achieving students.

[92] Robert D. Atkinson and Merrilea Mayo, *Refueling the U.S. Innovation Economy: Fresh Approaches to STEM Education*, Information Technology and Innovation Forum, December 7, 2010.

[93] Eric W. Robelen, "Latest Wave of STEM Schools Taps New Talent," *Education Week*, vol. 31, no. 3 (September 14, 2011), p.1, and published online under the title, "New STEM Schools Target Underrepresented Groups," *Edweek.org*.

[94] For example, the President's Council of Advisors on Science and Technology calculates that the United States will need, over the next decade, "approximately one million more college graduates in STEM fields than expected under current assumptions." Executive Office of the President, President's Council of Advisors on Science and Technology, *Engage to Excel: Producing One Million Additional College Graduates with Degrees in Science, Technology, Engineering, and Mathematics*, cover letter, February 2012.

[95] One-quarter of first-year college students were required to take remedial courses because they were not ready for college-level work. (Hart Research Associates, *One Year Out: Findings from a National Survey Among Members of the High School Graduating Class of 2010*, The College Board, August 18, 2011.) Two-fifths of students entering college intending to major in a STEM field complete a STEM degree. (Executive Office of the President, President's Council of Advisors on Science and Technology, *Engage to Excel: Producing One Million Additional College Graduates with Degrees in Science, Technology, Engineering, and Mathematics*, February 2012.)

[96] Council for Opportunity in Education, "2011 Capitol Hill Talking Points: Upward Bound Math-Science," talking points, 2011.

[97] Business Roundtable, *Taking Action for America: A CEO Plan for Jobs and Economic Growth*, March 1, 2012.

[98] Robin Wilson, "For-Profit Colleges Change Higher Education's Landscape," *The Chronicle of Higher Education*, February 7, 2010.

[99] For example, one widely cited study of STEM attrition found that poor teaching quality is a factor. (Elaine Seymour and Nancy M. Hewitt, *Talking About Leaving: Why Under-Graduates Leave the Sciences* (Boulder, CO: Westview Press, 1997). Other analysts cite the influence of grades on students' decisions to leave STEM majors. (Ben Ost, "The Role of Peers and Grades in Determining Major Persistence in the Sciences," *Economics of Education Review*, vol. 29, no. 6 (December 2010), pp. 923-934. Some observers assert that certain institutional practices, such as using introductory STEM courses to "weed out" or limit the number of students seeking STEM majors, contribute to perceived attrition challenges. (Jeffrey Mervis, "Weed-out Courses Hamper Diversity," *Science*, vol. 334, no. 6961 (December 2011), p. 1333.)

[100] Executive Office of the President, President's Council of Advisors on Science and Technology, *Engage to Excel: Producing One Million Additional College Graduates with Degrees in Science, Technology, Engineering, and Mathematics*, February 2012.

[101] For example, see National Academy of Sciences, National Academy of Engineering, and Institute of Medicine; Committee on Underrepresented Groups and the Expansion of the Science and Engineering Workforce Pipeline; Committee on Science, Engineering, and Public Policy; Policy and Global Affairs, *Expanding Underrepresented Minority Participation: America's Science and Technology Talent at the Crossroads* (Washington, DC: National Academies Press, 2011).

[102] Based on 2009 enrollment in four-year, two-year, and less-than-two-year Title VI eligible institutions. U.S. Department of Education, Institute of Education Sciences, National Center for Education Statistics, *Digest of Education Statistics: Table 196*, May 2011.

[103] Generally, analysts consider a demographic group to be "underrepresented" in STEM if the group's rate of participation in the STEM field is inconsistent with the group's presence in some broader population. For example, if women make up over half of all college students but are only a quarter of the engineering majors, then some observers would consider women to be underrepresented in college engineering enrollments.

[104] National Academy of Sciences, National Academy of Engineering, and Institute of Medicine; Committee on Underrepresented Groups and the Expansion of the Science and Engineering Workforce Pipeline; Committee on Science, Engineering, and Public Policy; Policy and Global Affairs, *Expanding Underrepresented Minority Participation: America's Science and Technology Talent at the Crossroads* (Washington, DC: National Academies Press, 2011), pp. 1-2. Also, although not specific to STEM, one 2009 report found that the U.S. Gross Domestic Product (GDP) could have been between two to four percent higher if the achievement gap between Latino/black and white students were narrowed. McKinsey & Company, Social Sector Office, *The Economic Impact of the Achievement Gap in America's Schools*, April 2009.

[105] For example, see David Beede et al., *Education Supports Racial and Ethnic Equality in STEM,* U.S. Department of Commerce, Economics and Statistics Administration, ESA Issue Brief #05-11, September 2011; Irwin Kirsch et al., *America's Perfect Storm: Three Forces Changing Our Nation's Future*, Educational Testing Service, Policy Information Center, Policy Evaluation and Research Center, January 2007; and National Academy of Sciences, National Academy of Engineering, and Institute of Medicine; Committee on Underrepresented Groups and the Expansion of the Science and Engineering Workforce Pipeline; Committee on Science, Engineering, and Public Policy; Policy and Global Affairs,

Expanding Underrepresented Minority Participation: America's Science and Technology Talent at the Crossroads (Washington, DC: National Academies Press, 2011).

[106] Each of the terms and factors from the 2011 review (as summarized in this report) are described in greater detail in Samuel D. Museus et al., "Racial and Ethnic Minority Students' Success in STEM Education," *ASHE Higher Education Report,* vol. 36, no. 6 (January 2011).

[107] The term "self-efficacy" refers to a student's confidence in his or her ability to learn STEM subjects.

[108] Samuel D. Museus et al., "Racial and Ethnic Minority Students' Success in STEM Education," *ASHE Higher Education Report,* vol. 36, no. 6 (January 2011), p. viii.

[109] "Stereotype threat" is a theory developed by some social psychologists to explain the perceived effects of negative group stereotypes on the academic performance of the targets of those stereotypes. In essence, the theory asserts that when confronted with negative group stereotypes (e.g., girls are bad at math), individuals perform more poorly than they are otherwise capable of doing on intellectual tests. (Claude M. Steele and Joshua Aronson, "Stereotype Threat and the Intellectual Test Performance of African Americans," *Journal of Personality and Social Psychology*, vol. 69. no. 5 (1995), pp. 797-811.)

[110] "Oppositional culture" is a theory developed by some social scientists to explain the academic disengagement of black students. In essence, the theory postulates that black Americans have formed a culture that opposes mainstream values (as reaction to racial oppression and discrimination) and that this oppositional culture leads black Americans to devalue academic success because it is associated with "acting white." (Signithia Fordham and John Ogbu, "Black Students' School Success: Coping with the Burden of Acting White," *Urban Review*, vol. 18, no. 3 (December 1985), pp. 176-206.) Some researchers criticize oppositional culture theory, arguing that African Americans actually maintain more pro-school values than whites, but that they lack the material conditions that foster the development of skills, habits, and styles rewarded by teachers. (James W. Ainsworth-Darnell and Douglas B. Downey, "Assessing the Oppositional Culture Explanation for Racial/Ethnic Differences in School Performance," *American Sociological Review*, vol. 63, no. 4 (August 1998), pp. 536-553.) In reflecting on the debate about oppositional culture theory, the authors of the 2011 review conclude, "this theory could plausibly be used to explain the negative educational outcomes of racial and ethnic minorities in K-12, particularly in STEM education." (Museus et. al., p. 37.)

[111] Samuel D. Museus et al., "Racial and Ethnic Minority Students' Success in STEM Education," *ASHE Higher Education Report,* vol. 36, no. 6 (January 2011), pp. vii-viii.

[112] Samuel D. Museus et al., "Racial and Ethnic Minority Students' Success in STEM Education," *ASHE Higher Education Report,* vol. 36, no. 6 (January 2011), p. viii.

[113] James J. Heckman, *The American Family in Black and White: A Post-Racial Strategy for Improving Skills to Promote Equality*, National Bureau of Economic Research, Working Paper no. 16841, March 2011.

[114] Irwin Kirsch et al., *America's Perfect Storm: Three Forces Changing Our Nation's Future*, Educational Testing Service, Policy Information Center, Policy Evaluation and Research Center, January 2007.

[115] Educational Testing Service, Policy Evaluation and Research Center, *Positioning Young Black Boys for Educational Success*, Policy Notes: News from the ETS Policy Information Center, vol. 19, no. 3 (Fall 2011); and Eugene E. Garcia et al., "Early Academic Achievement of Hispanics in the United States: Implications for Teacher Preparation," *New Educator*, vol. 2, no. 2 (2006), pp. 123-147.

[116] Kevin Fiscella and Harriet Kitzman, "Disparities in Academic Achievement and Health: The Intersection of Child Education and Health Policy," *Pediatrics*, vol. 123, no. 3 (March 2009).

[117] P.L. 110-69, Section 7032.

[118] National Academy of Sciences, National Academy of Engineering, and Institute of Medicine; Committee on Underrepresented Groups and the Expansion of the Science and Engineering Workforce Pipeline; Committee on Science, Engineering, and Public Policy; Policy and Global Affairs, *Expanding Underrepresented Minority Participation: America's Science and Technology Talent at the Crossroads* (Washington, DC: National Academies Press, 2011).

[119] Matthew Ladner and Lindsey Burke, *Closing the Racial Achievement Gap: Learning from Florida's Reforms*, The Heritage Foundation, Backgrounder no. 2468, October 4, 2010.

[120] The White House Domestic Policy Council and U.S. Department of Education, *Preserving a Critical National Asset: America's Disadvantaged Students and the Crisis in Faith-Based Urban Schools*, September 2008, p. 7.

[121] Alicia C. Dowd, Lindsey E. Malcolm, and Elsa E. Macias, *Improving Transfer Access to STEM Bachelor's Degrees at Hispanic Serving Institutions through the America COMPETES Act*, University of Southern California, Rossier School of Education, Center for Urban Education, March 2010; and Steve Olsen and Jay B. Labov, rapporteurs, *Community Colleges in the Evolving STEM Education Landscape*, National Research Council and National Academy of Engineering; Division on Policy and Global Affairs, Board on Higher Education and Workforce; Division on Earth and Life Studies, Board on Life Sciences; Division on Behavioral and Social Sciences and Education, Board on Science Education; National Academy of Engineering, Engineering Education Program Office; and Division on Behavioral and Social Sciences and Education, Teacher Advisory Council; Planning Committee on Evolving Relationships and Dynamics Between Two- and Four-Year Colleges, and Universities (Washington, DC: National Academies Press, May 8, 2012); and Alicia C. Dowd and Lindsey E. Malcom, *Reducing Undergraduate Debt to Increase Latina and Latino Participation in STEM Professions*, University of Southern California, Rossier School of Education, Center for Urban Education, May 2012.

[122] Gina A. Garcia and Sylvia Hurtado, *Predicting Latina/o STEM Persistence at HSIs and Non-HSIs*, University of California, Los Angeles, Graduate School of Education and Information Studies, Higher Education Research Institute, April 2011.

[123] For example, a March 16, 2010, congressional hearing on broadening participation in STEM included testimony on the role of MSIs in producing minority STEM graduates. U.S. Congress, House Committee on Science and Technology, Subcommittee on Research and Science Education, *Broadening Participation in STEM*, hearings, 111th Cong., 2nd sess., Serial No. 111-85 (Washington, DC: GPO, 2010).

[124] U.S. Department of Commerce, Economics and Statistics Administration, *Women in STEM: A Gender Gap to Innovation*, ESA Issue Brief #04-11, August 2011.

[125] Catherine Hill, Christianne Corbett, and Andresse St. Rose, *Why So Few: Women in Science, Technology, Engineering, and Mathematics*, American Association of University Women, February 2010; and National Academy of Sciences, National Academy of Engineering, and Institute of Medicine, Committee on Maximizing the Potential of Women in Academic Science and Engineering, *Beyond Bias and Barriers: Fulfilling the Potential of Women in Academic Science and Engineering* (Washington, DC: National Academies Press, 2007).

[126] Stephen J. Ceci and Wendy M. Williams, "Understanding Current Causes of Women's Underrepresentation in Science," *Proceedings of the National Academy of Sciences*, vol. 108, no. 8 (February 7, 2011), p. 3157-3162.

[127] Corinne A. Moss-Racusin, et al, "Science Faculty's Subtle Gender Biases Favor Male Students," *Proceedings of the National Academy of Sciences*, vol. 109, no. 41 (October 9, 2012), p. 16474-16479; and Catherine Hill, Christianne Corbett, and Andresse St. Rose, *Why So Few: Women in Science, Technology, Engineering, and Mathematics*, American Association of University Women, February 2010, pp. 90-96.

[128] Stephen J. Ceci and Wendy M. Williams, "Understanding Current Causes of Women's Underrepresentation in Science," *Proceedings of the National Academy of Sciences*, vol. 108, no. 8 (February 7, 2011), p. 3161.

[129] Sean Reardon, "The Widening Achievement Gap Between the Rich and the Poor: New Evidence and Possible Explanations," *Whither Opportunity*, ed. Greg J. Duncan and Richard J. Murnane (New York, NY: Russell Sage Foundation, September 2011), p. 91.

[130] Other factors associated with summer learning loss include demographic characteristics. For example, one 2006 study found that high-performing African-American and Hispanic students lost more achievement than their European-American peers over the summer and that low-performing African-American and Hispanic students grow less than low-performing students in all groups. See Martha S. McCall et. al., *Achievement Gaps: An Examination of Differences in Student Achievement and Growth*, Northwest Evaluation Association, November 2006.

[131] Jennifer Sloan McCombs et al., *Making Summer Count: How Summer Programs Can Boost Children's Learning*, RAND Corporation, 2011.

[132] Suzanne E. Graham and Lauren E. Provost, *Mathematics Achievement Gaps Between Suburban Students and Their Rural and Urban Peers Increase Over Time*, University of New Hampshire, Carsey Institute, Issue Brief No. 52, Summer 2012.

[133] The Department of Health and Human Services (HHS) traces its history to establishment of the federal Marine Hospital Service, forerunner of the contemporary U.S. Public Health Service, in 1798. More information about HHS history is available at http://www.hhs.gov/about/hhshist.html.

[134] Unless otherwise indicated, historical STEM education measures in this section are described *as originally passed.* Most of these measures have been amended, in some cases quite significantly (including repeal), since they became law.

[135] John C. Fitzpatrick, ed., *Journals of the Continental Congress 1774-1789*, edited from the original records in the Library of Congress, vol. 28 (Washington, DC: GPO, 1933), p. 375, http://memory.loc.gov/cgi-bin/query/r?ammem/ hlaw:@field%28DOCID+@lit%28jc0281%29%29.

[136] Library of Congress, *Primary Documents in America History: Northwest Ordinance*, Library of Congress/Virtual Services Digital Reference Section/Web Guides website, April 25, 2012, http://www.loc.gov/rr/program/bib/ourdocs/ northwest.html.

[137] Library of Congress, *An Ordinance for the Government of the Territory of the United States, North-West of the River Ohio*, Library of Congress/American Memory/Documents from the Continental Congress and Constitutional Convention 1774-1789 website, no date, http://memory.loc.gov/cgi-bin/query/r?ammem/ bdsdcc:@field(DOCID+@lit(bdsdcc22501)).

[138] The statute establishing the MHS does not include a formal title for the act. For the sake of consistency with other headings in this section, CRS used the title "Marine Hospital Service Act of 1789" to describe 1 Stat. 605. A copy of this statute is available at

http://memory.loc.gov/cgi-bin/ampage?collId=llsl&file Name=001/llsl001.db&recNum=728.

[139] U.S Congress, Senate Committee on Labor and Public Welfare, *National Research Service Award Act*, report to accompany H.R. 7724, 93rd Cong., 1st sess., S. Rept. 93-381.

[140] National Archives and Records Administration, *Morrill Act (1862)*, National Archives and Records Administration/100 Milestone Documents website, no date, http://www.ourdocuments.gov/doc.php?doc=33, http://www.ourdocuments.gov/doc.php?doc=33.

[141] The 1890 measure required states that accepted funds to either (a) discount race in admissions, or (b) provide separate colleges for white and black students.

[142] Agriculture and Food Act of 1981 (P.L. 97-98).

[143] Library of Congress, *Primary Documents in America History: Morrill Act*, Library of Congress/Virtual Services Digital Reference Section/Web Guides website, July 30, 2010, http://www.loc.gov/rr/program /bib/ourdocs/ Morrill.html.

[144] For more information on STEM education at the NSF, see CRS Report R42470, *An Analysis of STEM Education Funding at the NSF: Trends and Policy Discussion*, by Heather B. Gonzalez.

[145] National Science Foundation Act of 1950 (P.L. 81-507), http://www.nsf.gov /about/history/legislation.pdf.

[146] Controversy over the founding of the NSF focused mostly on organizational questions, concerns about patents, and on other issues not related to STEM education.

[147] Dorothy Schaffter, *The National Science Foundation* (New York, NY: Frederick A. Praeger, Publishers, 1969), p. 96.

[148] National Defense Education Act (P.L. 85-864).

[149] For example, see Barbara Barksdale Clowse, *Brainpower for the Cold War: The Sputnik Crisis and National Defense Education Act of 1958* (Westport, CT: Greenwood Press, 1981), p. 147; and Wayne J. Urban, *More Than Science and Sputnik: The National Defense Education Act of 1958* (Tuscaloosa, AL: University of Alabama Press, 2010), p. 202.

[150] For more information on ESEA, see CRS Report RL33960, *The Elementary and Secondary Education Act, as Amended by the No Child Left Behind Act: A Primer*, by Rebecca R. Skinner.

[151] Particularly Title I, Part A, Program of Education for the Disadvantaged.

[152] NSF hosts a companion program that is also called Mathematics and Science Partnerships (MSP). As currently authorized, the two programs were designed to complement each other.

[153] Department of Education Organization Act (P.L. 96-88), http://www.eric.ed.gov /PDFS/ED180121.pdf.

[154] U.S. Congress, Senate Committee on Governmental Affairs, *Legislative History of P.L. 96-88, Department of Education Organization Act* (Part 2), committee print, 96th Cong., 2nd sess. (Washington, DC: GPO 1980), p. 1758.

[155] For more information about the America COMPETES Act see CRS Report R41819, *Reauthorization of the America COMPETES Act: Selected Policy Provisions, Funding, and Implementation Issues*, by Heather B. Gonzalez.

In: Science, Technology, Engineering, … ISBN: 978-1-63463-126-6
Editor: Timothy Curtis

Chapter 2

SCIENCE, TECHNOLOGY, ENGINEERING, AND MATHEMATICS EDUCATION: ASSESSING THE RELATIONSHIP BETWEEN EDUCATION AND THE WORKFORCE*

United States Government Accountability Office

WHY GAO DID THIS STUDY

Federal STEM education programs help enhance the nation's global competitiveness by preparing students for STEM careers. Researchers disagree about whether there are enough STEM workers to meet employer demand. GAO was asked to study the extent to which STEM education programs are aligned with workforce needs.

GAO examined (1) recent trends in the number of degrees and jobs in STEM fields, (2) the extent to which federal postsecondary STEM education programs take workforce needs into consideration, and (3) the extent to which federal K-12 STEM education programs prepare students for postsecondary STEM education. GAO analyzed trends in STEM degrees and jobs since 2002 using 3 data sets—the Integrated Postsecondary Education Data System, American Community Survey, and Occupational Employment Statistics—and

* This is an edited, reformatted and augmented version of the United States Government Accountability Office publication, GAO-14-374, dated May 2014.

surveyed 158 federal STEM education programs. There were 154 survey respondents (97 percent): 124 postsecondary and 30 K-12 programs. In addition, GAO conducted in-depth reviews—including interviews with federal officials and grantees—of 13 programs chosen from among those with the highest reported obligations.

What GAO Recommends

GAO makes no recommendations in this report. GAO received technical comments from the Departments of Education, Energy, and Health and Human Services; National Science Foundation; and Office of Management and Budget.

What GAO Found

Both the number of science, technology, engineering, and mathematics (STEM) degrees awarded and the number of jobs in STEM fields increased in recent years. The number of degrees awarded in STEM fields grew 55 percent from 1.35 million in the 2002-2003 academic year to over 2 million in the 2011-2012 academic year, while degrees awarded in non-STEM fields increased 37 percent. Since 2004, the number of STEM jobs increased 16 percent from 14.2 million to 16.5 million jobs in 2012, and non-STEM jobs remained fairly steady. The trends in STEM degrees and jobs varied across STEM fields. It is difficult to know if the numbers of STEM graduates are aligned with workforce needs, in part because demand for STEM workers fluctuates. For example, the number of jobs in core STEM fields, including engineering and information technology, declined during the recession but has grown substantially since then.

Almost all of the 124 federal postsecondary STEM education programs that responded to GAO's survey reported that they considered workforce needs in some way. For example, the most common program objective was to prepare students for STEM careers. Some of these programs focused on occupations they considered to be in demand and/or related to their agency's mission. Many postsecondary programs also aimed to increase the diversity of the STEM workforce or prepare students for innovation. Most STEM programs reported having some outcome measures in place, but GAO found that some programs did not measure an outcome directly related to their stated objectives. As GAO recommended in 2012, the National Science and Technology Council recently issued guidance to help agencies better incorporate STEM education outcomes into their performance plans and reports. As agencies follow the guidance and focus on the effectiveness of the programs, more programs may measure outcomes directly related to their objectives. Of the 30 kindergarten through 12th grade (K-12) STEM education programs responding to GAO's survey, almost all reported that they either directly or indirectly prepared students for postsecondary STEM education. For example, one program worked closely with students to provide math and science instruction and supportive services to prepare them for postsecondary STEM education, while another supported research projects intended to enhance STEM learning.

ABBREVIATIONS

ACS	American Community Survey
CIP	Classification of Instructional Program
IPEDS	Integrated Postsecondary Education Data System
IT	Information technology
K-12	Kindergarten-12th grade
OES	Occupational Employment Statistics
SOC	Standard Occupational Classification
STEM	Science, technology, engineering, and mathematics

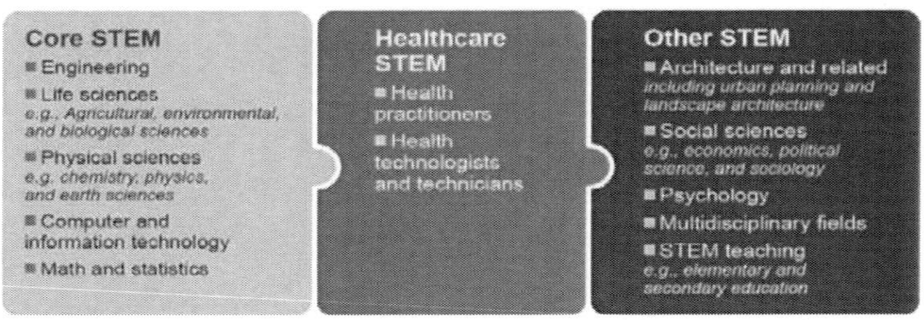

Source: GAO categories based on Options for Defining STEM Occupations Under the 2010 Standard Occupational Classification (SOC) System: SOC Policy Committee Recommendation to the Office of Management and Budget (August 2012).

Science, Technology, Engineering, and Mathematics (STEM) Fields.

May 8, 2014
The Honorable John Kline
Chairman
Committee on Education and the Workforce
House of Representatives

The Honorable Richard Hanna
House of Representatives

The Honorable Joseph Heck
House of Representatives

The Honorable Duncan D. Hunter
House of Representatives

Science, technology, engineering, and mathematics (STEM) education programs can serve an important role both by helping to prepare students and teachers for careers in STEM fields and by enhancing the nation's global competitiveness. As part of this effort, many federal agencies administer STEM education programs. In addition to the federal effort, state and local governments, universities and colleges, and the private sector have also developed programs that provide opportunities for students to pursue STEM education and occupations.

The current administration maintains that a strong educational pipeline creating future STEM workers is important to ensure that the United States remains competitive with other highly technological nations. Researchers disagree about the sufficiency of the current supply of STEM workers. While some researchers have concluded that the United States has a sufficient supply of STEM workers,[1] others have found that the educational system is not producing enough STEM graduates to fill the jobs available in STEM occupations[2] or in the increasing number of jobs in other fields that may require STEM competencies (such as analytical skills).[3] In light of this disagreement, we were asked to review the alignment between STEM and workforce needs. Specifically, we reviewed (1) recent trends in the number of degrees and jobs in STEM fields, (2) the extent to which federal postsecondary STEM education programs take workforce needs into consideration, and (3) the extent to which federal kindergarten-12th grade (K-12) STEM education programs prepare students for postsecondary STEM education.

To address our first objective, we analyzed three federal data sources to examine trends in STEM degrees and jobs over the past decade: (1) the Department of Education's Integrated Postsecondary Education Data System to examine trends in STEM degrees, (2) the Bureau of Labor Statistics' Occupational Employment Statistics data to examine employment and wage trends among STEM workers, and (3) the Census Bureau's American Community Survey data to examine unemployment rates of STEM workers and the relationships between educational background and occupation among STEM workers. We determined that these data were sufficiently reliable for the purposes of our report by reviewing relevant documentation and conducting electronic testing of the data. We also conducted a regression analysis with the American Community Survey data to examine differences in wages and unemployment rates between STEM and non-STEM workers, controlling for some demographic details.

To address our other objectives, we reviewed relevant federal laws and regulations and conducted and analyzed the results of a survey. We surveyed 158 K-12 and postsecondary STEM education programs about how they address STEM workforce needs and prepare students for future STEM education or careers. A total of 154 federal STEM education programs responded to our survey, representing a 97 percent response rate.[4] Among the respondents, we identified 124 programs—reporting $1.9 billion in fiscal year 2012 obligations—that primarily served students and teachers at the postsecondary level, and 30 programs—reporting $685 million in fiscal year 2012 obligations—that primarily served students and teachers at the K-12 level.[5] To provide more details about some of the highest funded STEM education programs, we conducted a more in-depth review of 13 programs from three agencies: the National Science Foundation, the Department of Education, and the National Institutes of Health at the Department of Health and Human Services. We chose these programs because they were among the largest federal STEM education programs, collectively accounting for 54 percent of the fiscal year 2012 STEM education obligations reported by the 154 programs that responded to our survey. Seven of the selected programs served postsecondary students or institutions and six programs served K-12 students or teachers. We reviewed documentation from each program, interviewed agency officials, and conducted site visits with program grantees in Austin, Texas, and San Francisco, California, and phone interviews with grantees in Boston, Massachusetts.

We conducted this performance audit from January 2013 to April 2014 in accordance with generally accepted government auditing standards. Those standards require that we plan and perform the audit to obtain sufficient, appropriate evidence to provide a reasonable basis for our findings and conclusions based on our audit objectives. We believe that the evidence obtained provides a reasonable basis for our findings and conclusions based on our audit objectives.

BACKGROUND

STEM Education Definitions

The term "STEM education" refers to teaching and learning in the fields of science, technology, engineering, and mathematics. It includes educational activities across all grade levels—from pre-school to post-doctorate—in both formal (e.g., classrooms) and informal (e.g., afterschool programs) settings.[6]

- In 2012, we reviewed the delivery and effectiveness of federal STEM education programs. As in our 2012 report, for this report we define a federally-funded STEM education program as a program funded in a designated fiscal year[7] by allocation or congressional appropriation that includes one or more of the following as a primary objective:[8]
- attract or prepare students to pursue classes or coursework in STEM areas through formal or informal education activities,
- attract students to pursue degrees (2-year, 4-year, graduate, or doctoral) in STEM fields through formal or informal education activities,
- provide training opportunities for undergraduate or graduate students in STEM fields (this can include grants, fellowships, internships, and traineeships that are targeted to students; we do not consider general research grants to researchers that may hire a student to work in the lab to be a STEM education program),
- attract graduates to pursue careers in STEM fields,
- improve teacher education in STEM areas for teachers and those studying to be teachers,
- improve or expand the capacity of K-12 schools or postsecondary institutions to promote or foster education in STEM fields, or
- conduct research to enhance the quality of STEM education programs provided to students.

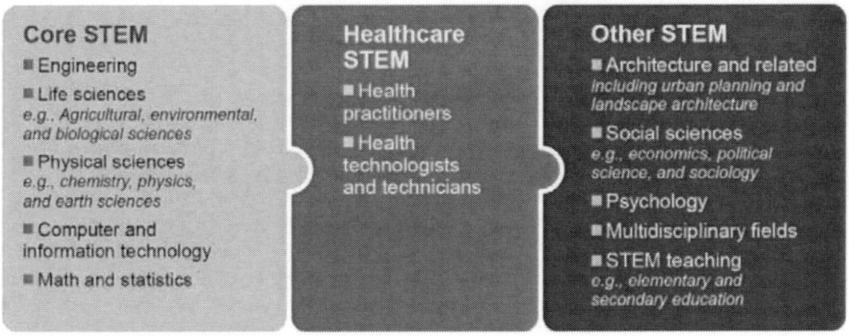

Source: GAO categories based on Options for Defining STEM Occupations Under the 2010 Standard Occupational Classification (SOC) System: SOC Policy Committee Recommendation to the Office of Management and Budget (August 2012).

Figure 1. Science, Technology, Engineering, and Mathematics (STEM) Fields.

There is no commonly used definition of fields that are considered STEM. For this report, we use a comprehensive definition of STEM that includes three STEM categories: Core STEM, Healthcare STEM, and Other STEM (see figure 1).[9] We present our findings for the three categories combined and for each of three STEM categories. See our description of the relevant data sets in appendix I for an explanation of how we classified fields of study and occupations into these STEM categories in our data analysis.

Federal STEM Education Programs and Policy

The Committee on STEM Education is the interagency coordination body for STEM education in the federal government (see figure 2).

Federal STEM education programs have been created in two ways—directly by law or through agencies' broad statutory authority to carry out their missions. In our 2012 STEM report,[10] we reported that in fiscal year 2010, 13 federal agencies administered 209 programs to increase knowledge of STEM fields and attainment of STEM degrees. These agencies, listed below in Table 1, continued to administer federal STEM education programs in fiscal year 2014.

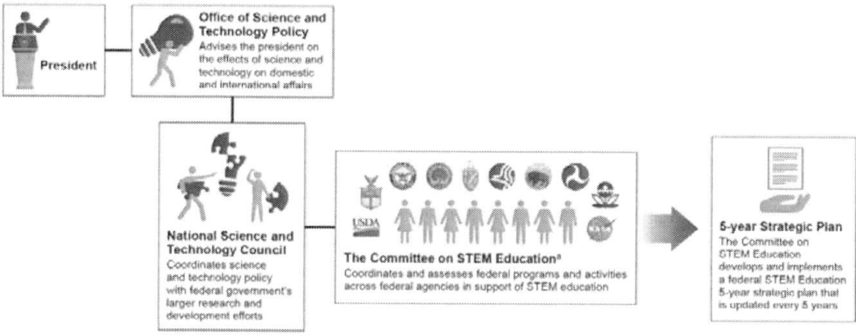

Source: GAO.

[a] The Committee on STEM Education coordinates federal programs and activities in support of STEM education, as required by the America COMPETES Reauthorization Act of 2010 (Pub. L. No. 111-358, § 101(a), 124 Stat. 3982, 3984 (2011)). The Act also approved funding for some STEM education programs and addressed coordination and oversight issues, including those associated with the coordination and potential duplication of federal STEM education efforts. Specifically, the Act required the Director of Office of Science and Technology Policy to establish a committee under the National Science and Technology Council to inventory, review, and coordinate federal STEM education programs, among other things.

Figure 2. Science, Technology, Engineering, and Mathematics (STEM) Education Policy.

In our 2012 report, we found that in fiscal year 2010, 83 percent of the programs we identified overlapped to some degree with at least 1 other program by offering similar services to similar target groups in similar STEM fields to achieve similar objectives. Although those programs may not be duplicative, we reported that they were similar enough that they needed to be well coordinated and guided by a robust strategic plan. We also found that federal agencies' limited use of performance measures and evaluations may have hampered their ability to assess the effectiveness of individual programs as well as the overall federal STEM education effort. We recommended that as the Office of Science and Technology Policy leads the government's STEM education strategic planning effort, it should work with agencies to better align their activities with a government-wide strategy, develop a plan for sustained coordination, identify programs for potential consolidation or elimination, and assist agencies in determining how to better evaluate their programs.

Table 1. Agencies Administering Science, Technology, Engineering, and Mathematics (STEM) Education Programs

Agency	Mission
Department of Agriculture	To provide leadership on food, agriculture, natural resources, and related issues based on sound public policy, the best available science, and efficient management
Department of Commerce	To promote job creation, economic growth, sustainable development, and improved standards of living for all Americans by working in partnership with businesses, universities, communities, and our nation's workers
Department of Defense	To provide the military forces needed to deter war and to protect the security of our country
Department of Education	To promote student achievement and preparation for global competitiveness by fostering educational excellence and ensuring equal access
Department of Energy	To ensure America's security and prosperity by addressing its energy, environmental and nuclear challenges through transformative science and technology solutions
Department of Health and Human Services	To enhance the health and well-being of Americans by providing for effective health and human services and by fostering sound, sustained advances in the sciences underlying medicine, public health, and social services
Department of Homeland Security	To ensure a homeland that is safe, secure, and resilient against terrorism and other hazards
Department of the Interior	To protect and manage the nation's natural resources and cultural heritage; to provide scientific and other information about those resources; and to honor its trust responsibilities or special commitments to American Indians, Alaska Natives, and affiliated island communities
Department of Transportation	To ensure a fast, safe, efficient, accessible and convenient transportation system that meets our vital national interests and enhances the quality of life of the American people, today and into the future
Environmental Protection Agency	To protect human health and the environment
National Aeronautics and Space Administration	To drive advances in science, technology, and exploration to enhance knowledge, education, innovation, economic vitality, and stewardship of Earth

Table 1. (Continued)

Agency	Mission
National Science Foundation	To promote the progress of science; to advance the national health, prosperity, and welfare; to secure the national defense; and for other purposes
Nuclear Regulatory Commission	To ensure the adequate protection of public health, safety, and the environment while promoting the common defense and security

Source: GAO review of agencies' websites and strategic plans.

The Office of Science and Technology Policy has taken steps to address some of our recommendations. Regarding our recommendation on potential elimination or consolidation of programs, the Committee on STEM Education released its interim strategic planning progress report in February 2012, which noted that STEM education programs had been identified to be potentially overlapping and encouraged agencies to streamline programs where appropriate.[11] In addition, the President's fiscal year 2014 budget called for a major restructuring of federal STEM education programs through the consolidation of programs and the realignment of STEM education activities.[12] Since our prior report on STEM, the number of STEM education programs dropped from 209 in 2010 to158 in 2013. The President's fiscal year 2015 budget request seeks to continue these efforts and states that agencies should focus on internal consolidations and eliminations while funding their most effective programs.[13] Regarding our recommendation on evaluations, in May of 2013 the Committee on STEM Education released its 5-year Strategic Plan, which included guidance to agencies in developing evaluations for STEM education programs. The plan also laid out five broad priority areas:[14]

- Improve STEM instruction;
- Increase and sustain youth and public engagement in STEM;
- Enhance STEM experiences of undergraduate students;
- Better serve groups historically under-represented in STEM fields; and
- Design graduate education for tomorrow's STEM workforce.

In addition, in July 2013, a joint Office of Science and Technology Policy/ Office of Management and Budget memo included guidance to agencies on how to align their programs and budget submissions—beginning with the budget submission for 2015—with the goals of the STEM Education 5-Year

Strategic Plan. The guidance includes language directing the agencies to prioritize programs that use evidence to guide program design and implementation and to define appropriate metrics and improve the measurement of outcomes. Furthermore, in the President's 2015 budget submission, the administration stated that improving STEM education by implementing the 5-year Strategic Plan is a cross-agency priority goal. As a result of this designation, the Office of Management and Budget must review on a quarterly basis agencies' progress in meeting this goal.

STEM Degrees and Jobs Are Increasing, but Their Alignment Is Difficult to Determine

While Degrees Have Increased in Most STEM Fields, Some Fields Have Grown More than Others

Overall, postsecondary degrees awarded in STEM fields have increased at a greater rate than in non-STEM fields during the past decade.[15] The number of STEM degrees awarded increased 55 percent, from 1.35 million degrees awarded in the 2002-2003 academic year to over 2 million in the 2011-2012 academic year. In comparison, degrees awarded in non-STEM fields increased 37 percent in the same time period (see figure 3). STEM degrees now comprise a larger share of total postsecondary degrees awarded—42 percent in the 2011-2012 academic year, up from 39 percent in the 2002-2003 academic year.

However, much of the increase in STEM degrees came from growth in awards of Healthcare degrees, which have doubled over the past decade (see figure 4). Degrees awarded in Core STEM fields increased at a substantially lower rate (19 percent) than non-STEM fields (37 percent).

Degrees awarded in Other STEM fields increased at a greater rate (43 percent) than non-STEM fields.[16]

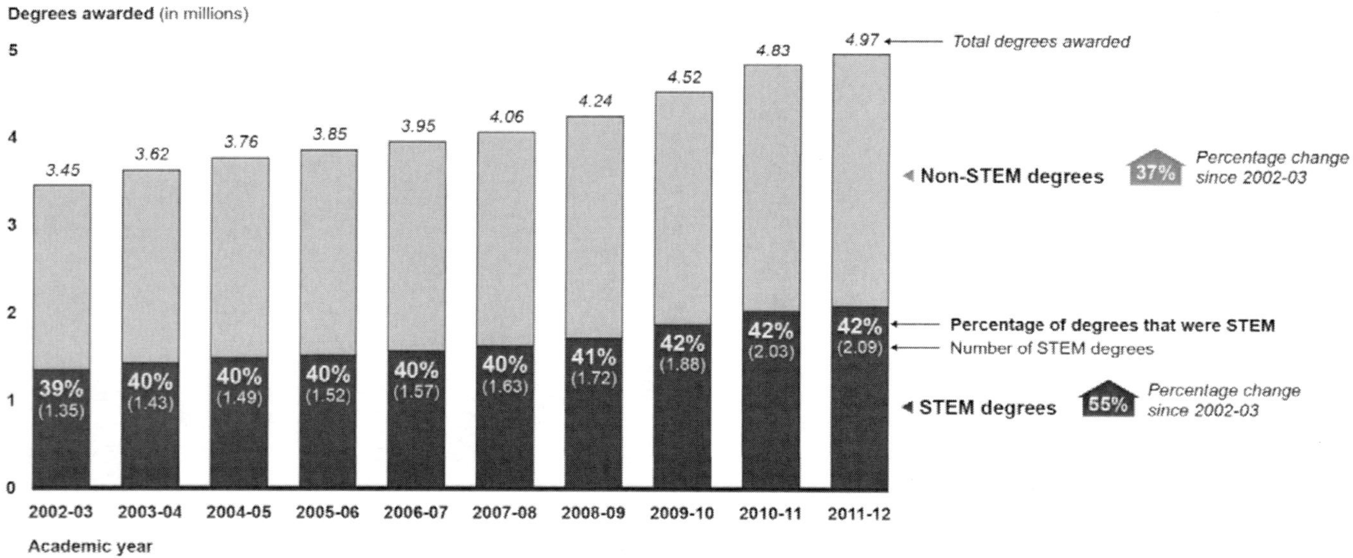

Source: GAO analysis of data from the Integrated Postsecondary Education Data System.

Note: This figure presents data on all degrees, including associate's and other degrees below the bachelor's level, bachelor's, master's, postbaccalaureate and post-master's certificates, doctorate, and professional degrees. The numbers shown in this figure include degrees awarded to citizens, non-resident aliens (students in the United States on a visa or temporary basis and do not have the right to remain indefinitely), and resident aliens.

Figure 3. Trends in Science, Technology, Engineering, and Mathematics (STEM) and Non-STEM Degrees Awarded.

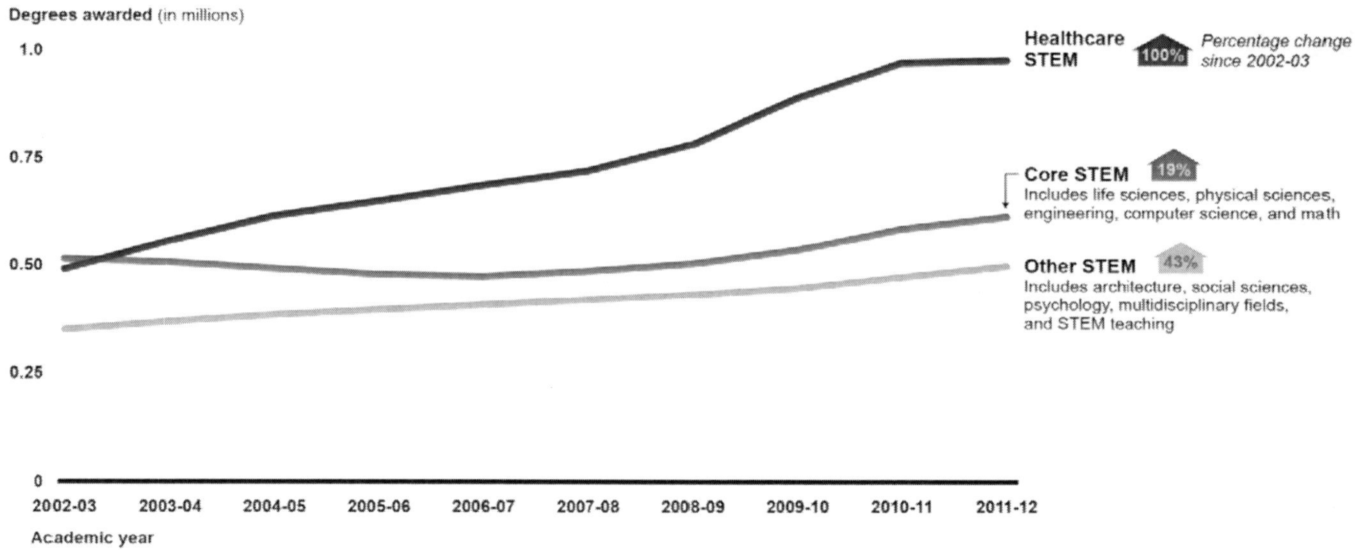

Source: GAO analysis of data from the Integrated Postsecondary Education Data System.

Note: This figure presents data on all degrees, including associate's and other degrees below the bachelor's level, bachelor's, master's, postbaccalaureate and post-master's certificates, doctorate, and professional degrees. The numbers shown in this figure include degrees awarded to citizens, non-resident aliens (students in the United States on a visa or temporary basis and do not have the right to remain indefinitely), and resident aliens.

Figure 4. Trends in Degrees Awarded in Science, Technology, Engineering, and Mathematics (STEM) Categories.

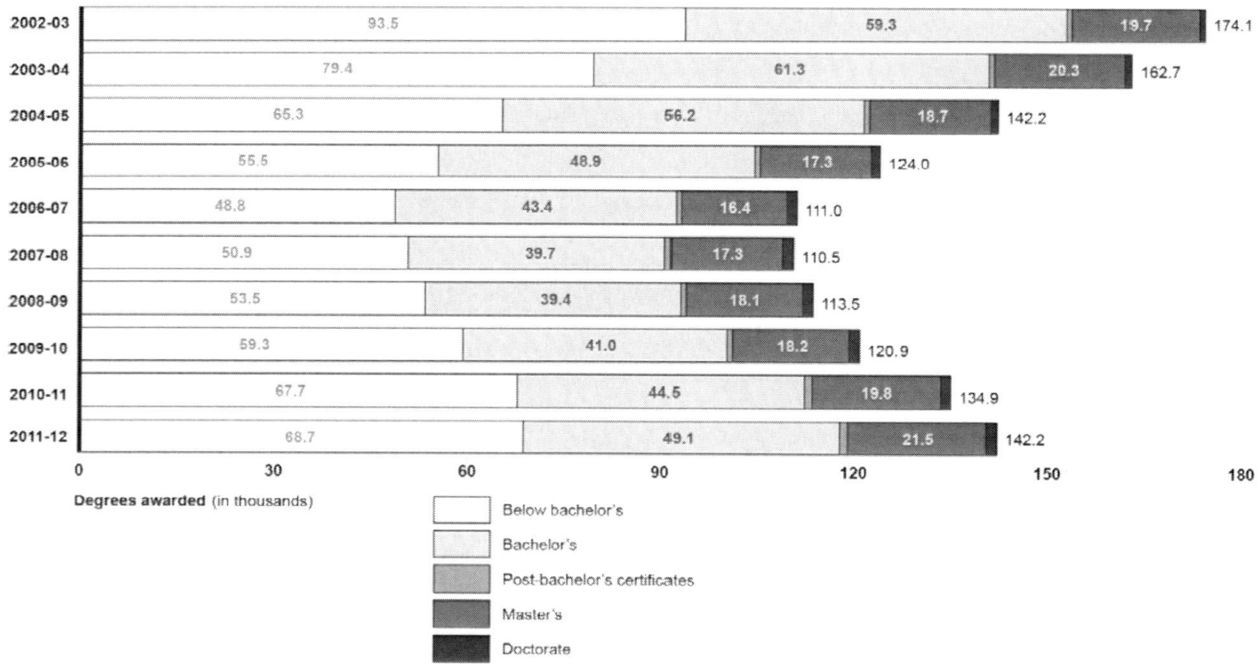

Source: GAO analysis of data from the Integrated Postsecondary Education Data System.

Note: Post-bachelor's certificates includes postbaccalaureate and post-master's certificates. The numbers shown in this figure include degrees awarded to citizens, non-resident aliens (students in the United States on a visa or temporary basis and do not have the right to remain indefinitely), and resident aliens.

Figure 5. Trends in Computer Science and Information Technology Degrees Awarded.

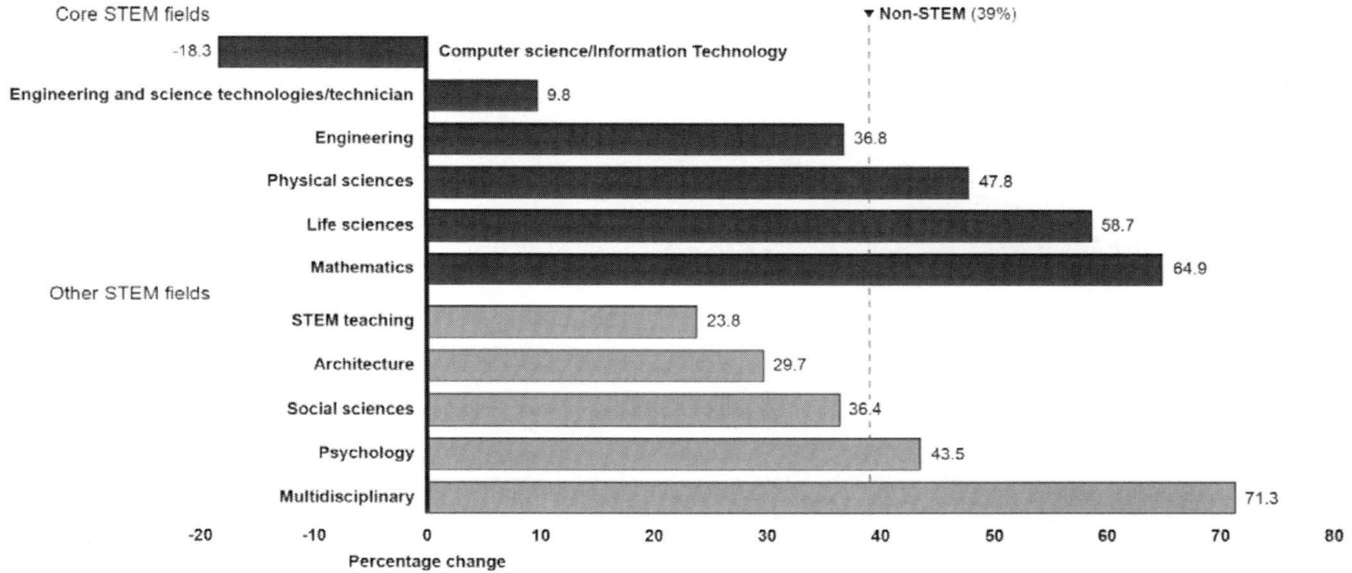

Source: GAO analysis of data from the Integrated Postsecondary Education Data System.

Note: This figure presents data on all degrees, including associate's and other degrees below the bachelor's level, bachelor's, master's, postbaccalaureate and post-master's certificates, doctorate, and professional degrees. The numbers shown in this figure include degrees awarded to citizens, non-resident aliens (students in the United States on a visa or temporary basis and do not have the right to remain indefinitely), and resident aliens.

Figure 6. Percentage Change in Postsecondary Degrees Awarded from the 2002-2003 to 2011-2012 Academic Years for Select Science, Technology, Engineering, and Mathematics (STEM) Fields.

The comparatively slower growth in Core STEM fields is due in large part to an 18 percent decline in the number of computer science and information technology (IT) degrees awarded in the past decade. Computer science and IT degrees decreased each year between the 2002-2003 and 2007-2008 academic years but then increased (see figure 5).[17] A research association that has examined trends in computer science bachelor's degrees attributes the decline to the "dot-com crash."[18] Aside from degrees awarded in the computer science/IT field, degrees awarded in all of the other STEM fields have increased throughout the past decade. Among the Core STEM fields, degrees awarded in the physical sciences, life sciences, and mathematics have grown at a greater rate than non-STEM fields (see figure 6). Degrees awarded in engineering have also increased, though at a slightly lower rate than non-STEM fields (37 percent compared to 39 percent).[19]

Employment Conditions Have been More Favorable in STEM Occupations than Non-STEM, though They Vary across STEM Fields

Overall, employment trends have generally been more favorable in STEM occupations than in non-STEM occupations. The number of jobs in STEM occupations increased 16 percent from 14.2 million jobs in 2004 to 16.5 million in 2012, while jobs in non-STEM occupations remained fairly steady (with a decline of 0.1 percent). STEM occupations also had more wage growth on average and lower unemployment rates than non-STEM occupations (see table 2). However, employment conditions vary across STEM fields, with healthcare occupations generally having the most favorable conditions. (See also appendix III for more detailed information on recent trends in STEM and non-STEM occupations).

Comparison of Workers in STEM Occupations to Workers with Similar Characteristics in Non-STEM Occupations

After controlling for education levels, demographic characteristics, and type of job, we estimate that the unemployment rate among workers in STEM occupations overall was 1.2 percentage points lower than for similar workers in non-STEM occupations in 2012, and the average wage in STEM occupations was 17 percent higher (see table 3). Healthcare occupations had the largest differences, while workers in Other STEM occupations had unemployment rates and average wages that were similar to those in non-STEM occupations.

Table 2. Trends in Science, Technology, Engineering, and Mathematics (STEM) and Non-STEM Occupations, 2004-2012

	Percent change, May 2004 to May 2012		
	Number of workers(2012 employment level)	**Average annual wage, inflation-adjusted (2012 mean wage)**	**Unemployment rate, 2012**
Non-STEM Occupations	- 0.1 (113.8 million workers)	+ 1.8 ($41,000)	8.4
STEM Occupations	+ 16 (16.5 million workers)	+ 6.2 ($79,000)	3.2
Core STEM	+ 11 (7.9 million workers)	+ 5.4 ($83,000)	3.9
Healthcare STEM	+ 21 (8.2 million workers)	+ 7.9 ($75,000)	2.6
Other STEM	+ 9 (0.5 million workers)	[a] ($74,000)	3.3

Source: GAO analysis of Occupational Employment Statistics data for the percentage change in number of workers and the average wage. GAO analysis of American Community Survey data for the unemployment rate in 2012.

[a] We do not report an estimate for Other STEM occupations because the margin of error at the 95 percent confidence level exceeds 30 percent of the estimate. The 95 percent confidence interval for the percentage change in average wage in other STEM occupations is 0.5 to 4.0 percent.

Note: With regard to the percentage change in number of workers between May 2004 and May 2012, the estimates shown in this table have margins of error within plus or minus 1.1 percentage points. Estimates of the number of workers employed in 2012 in each of the categories shown in the table have margins of error within plus or minus 1.5 percent of the estimate. With regard to the percentage change in average wage between May 2004 and May 2012, the estimates shown in this table have margins of error within plus or minus 0.8 percentage points. The estimates for the mean wage in 2012 have margins of error within plus or minus 2.1 percent of the estimate. With regard to the unemployment rates, estimates shown in this table have margins of error within plus or minus 0.2 percentage points, with the exception of other STEM occupations, which has a margin of error within plus or minus 0.6 percentage points. Differences between STEM and non-STEM occupations in the percentage change in employment level, percentage change in the average wage, and the unemployment rates are all statistically significant at the 95 percent confidence level. The differences between the three STEM categories in the percentage change in employment level and average wages were statistically significant at the 95 percent confidence level. With regard to unemployment rates, differences between healthcare STEM and the two other STEM categories were statistically significant at the 95 percent confidence level, but the difference between Core STEM and Other STEM was not statistically significant.

Table 3. Estimated Differences in Unemployment Rates and Annual Wages, Comparing Workers in Science, Technology, Engineering, and Mathematics (STEM) Occupations with Similar Workers in Non-STEM Occupations, 2012

Category	Estimated difference in unemployment rates, compared to similar workers in non-STEM occupations	Estimated difference in annual wages, compared to similar workers in non-STEM occupations
STEM Occupations	1.2 percentage points lower than non-STEM	17 percent higher than non-STEM
Core STEM	0.4 percentage points lower than non-STEM	16 percent higher than non-STEM
Healthcare STEM	2.0 percentage points lower than non-STEM	20 percent higher than non-STEM
Other STEM	similar to non-STEM	similar to non-STEM

Source: GAO analysis of American Community Survey data.

Note: Unemployment rate differences are measured in percentage points, while differences in wages are measured in percent of dollars. For example, a 2 percentage point increase in an unemployment rate of 10 percent would result in an unemployment rate of 12 percent. A 2 percent increase in a wage of $10 per hour would result in a wage of $10.20 per hour.

Employment Conditions across STEM Fields

While employment conditions have generally been more favorable in STEM occupations than in non-STEM occupations, conditions vary across specific STEM fields. Most STEM fields experienced both increases in employment levels and in average wages from 2004 to 2012, as well as relatively low unemployment rates when compared to non-STEM occupations. However, three fields— STEM sales occupations, engineering technician and drafting occupations, and science technician occupations—experienced either a decline in the number of jobs in this time period or a decline in the average wage (see figure 7). Engineering technician and drafting occupations and science technician occupations are also among the STEM fields with the highest unemployment rates in recent years, though their unemployment rates have fallen since 2010 and were lower than non-STEM occupations in 2012 (see figure 8).[20]

Several Factors, Including Fluctuations in the Economy, Make It Difficult to Determine Whether Supply of STEM Workers Is Aligned with Employer Demand

It is difficult to know whether the United States is producing enough STEM workers to meet employer needs for several reasons. First, estimating how many STEM workers employers need is a challenge, in part because demand for STEM workers can fluctuate with economic conditions. For example, the number of jobs in core STEM occupations declined by about 250,000 between 2008 and 2010 (from 7.74 million jobs in 2008 to 7.49 million in 2010), though it then increased (to 7.89 million jobs in 2012). Subject matter specialists and federal officials we interviewed also noted that employer needs in STEM fields are difficult to predict because they may change with technological or market developments.

Furthermore, the supply of STEM workers in the United States may not match the demand at any given point in time because of the time it takes to educate a STEM worker. Research suggests that students' decisions about which fields to study may be influenced by the economic conditions and future career prospects they perceive in those fields.[21] Favorable economic conditions in a STEM field may encourage students to pursue degrees in that field. However, it may take them several years to complete their degrees, so changes in the domestic supply of STEM workers may lag behind changes in the domestic demand.[22]

In addition, the number of students graduating with STEM degrees may not be a good measure of the supply of STEM workers because students often pursue careers in fields different from the ones they studied. Figure 9 shows the educational background of workers in selected STEM occupations in 2012 up to the bachelor's level. With the exception of engineering, most of those in STEM occupations did not receive a bachelor's degree in the same field in which they were working. They either majored in a different STEM field or a non-STEM field in their undergraduate education, or they did not receive a bachelor's degree.[23] As a result, it is difficult to estimate the supply of workers in a STEM occupation from information on the number of bachelor's degrees awarded in a STEM field.

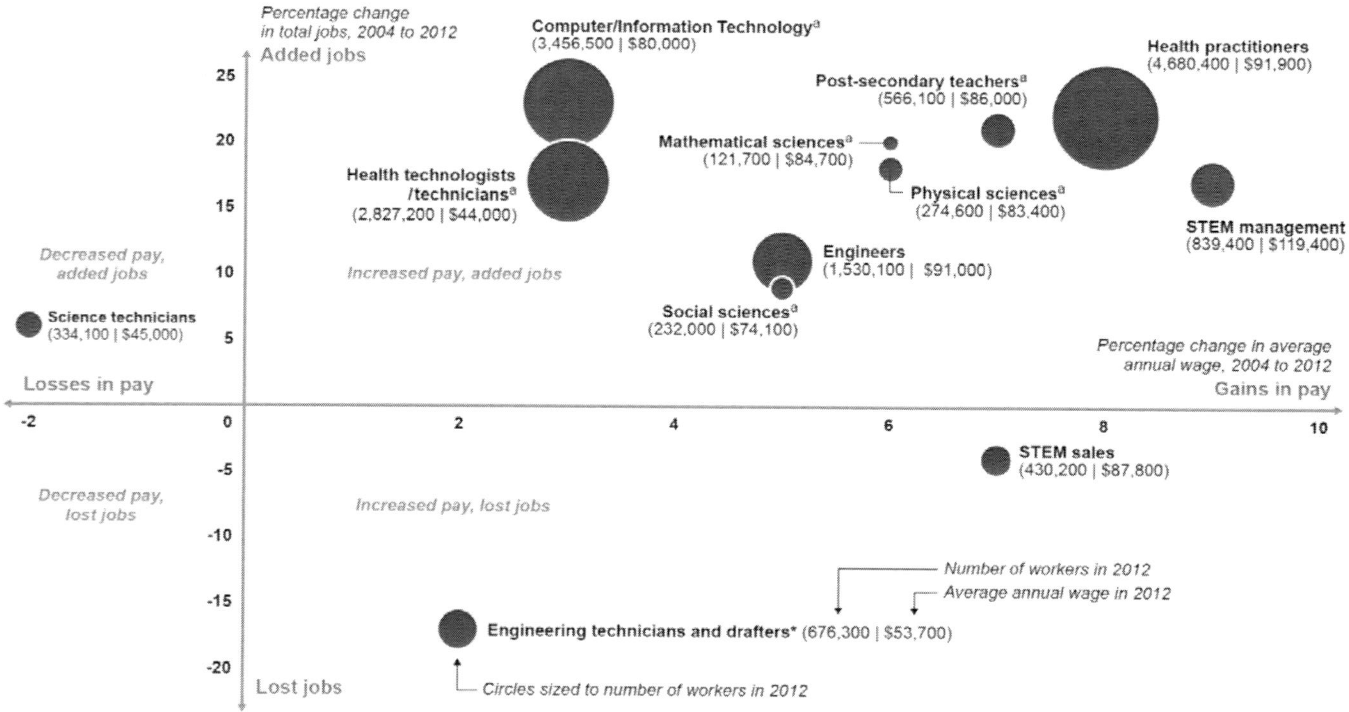

Source: GAO analysis of Occupational Employment Statistics data.

Note: Estimates of the percentage change in number of jobs between May 2004 and May 2012 have margins of error within plus or minus 3.1 percentage points. Estimates of the percentage change in average wages in this timeframe that do not have an asterisk have margins of error within plus or minus 1.9 percentage points. Estimates of the number of jobs in May 2012 have margins of error within plus or minus 2.7 percent of the estimate. Estimates of the average annual wage in May 2012 have margins of error within plus or minus 4.9 percent of the estimate.

[a] Indicates that the estimate of the percentage change in the average wage between May 2004 and May 2012 have margins of error greater than 30 percent of the estimate. The 95 percent confidence intervals for these estimates are: 1.2 to 3.9 percent increase in the average wage for computer science occupations, 3.5 to 7.9 percent for mathematical sciences occupations, 0.3 to 3.5 percent for engineering technicians and drafters, 4.0 to 8.2 percent for physical sciences occupations, 1.8 to 8.4 percent for social sciences occupations, 1.7 to 11.6 for postsecondary teachers, and 2.0 to 4.1 percent for health technologists and technicians.

Figure 7. Employment and Wage Trends across Science, Technology, Engineering, and Mathematics (STEM) Fields, 2004 to 2012.

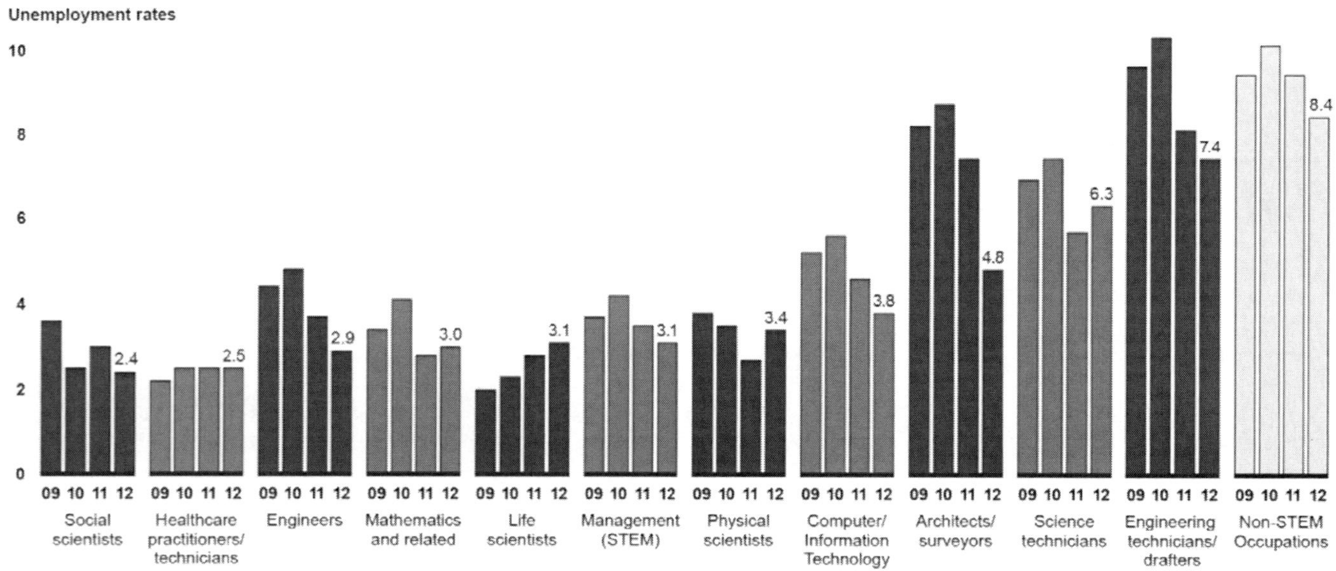

Source: GAO analysis of American Community Survey data.

Note: Estimates of the unemployment rate shown in this figure have margins of error within plus or minus 1.5 percentage points. For estimates that have a margin of error that exceeds 30 percent of the estimate, the 95 percent confidence intervals are: 1.3 to 2.7 for life sciences occupations in 2009, 1.6 to 3.0 for life sciences occupations in 2010, 1.3 to 2.5 for social science occupations in 2009, and 1.9 to 3.7 for mathematics occupations in 2011. Differences in the unemployment rates between STEM occupational groups and non-STEM occupations within each of the years are statistically significant at the 95 percent confidence level, except for the following: architects/surveyors in 2009 and engineering technicians/drafters in 2009 and 2010.

Figure 8. Unemployment Rates in Selected Science, Technology, Engineering, and Mathematics (STEM) Occupations, 2009-2012.

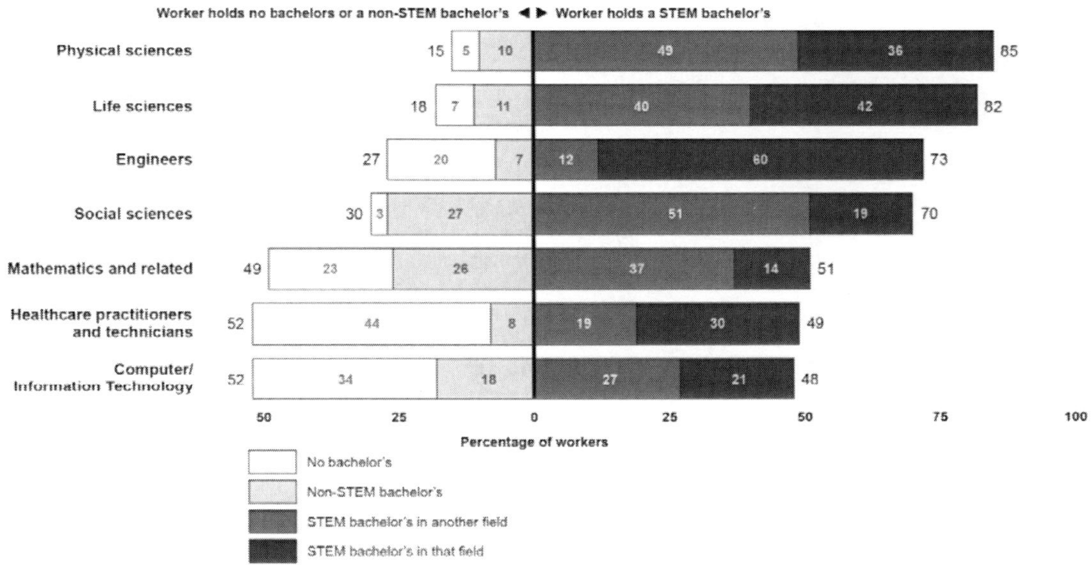

Source: GAO analysis of American Community Survey data.

Note: Estimates of the percentage of each occupational group in the education categories shown in this figure have margins of error that are within plus or minus 2.6 percentage points. In the estimates shown for engineers above, the percentages of workers with a STEM bachelor's degree in that field and a STEM bachelor's degree in another field does not sum to the numbers on the right of the bars due to rounding.

Figure 9. Educational Backgrounds of Workers Ages 22 or Older in Science, Technology, Engineering, and Mathematics (STEM) Occupations, 2012.

Further evidence of the difficulty in estimating the size of the STEM workforce from information on the number of STEM degrees is the substantial portion of workers with STEM bachelor's degrees who work in non-STEM occupations— 62 percent in 2012 (see figure 10). The survey data cannot tell us how many of these STEM-educated workers are in a non-STEM occupation by choice and how many would prefer to work in a STEM occupation but cannot find a position suitable to them. However, these workers have had relatively low unemployment rates in recent years— 4.8 percent in 2012, suggesting that there is generally demand in the workplace for workers with STEM education, both in STEM and non-STEM occupations (see appendix III for further information on the educational backgrounds of workers in STEM and non-STEM occupations).

MOST FEDERAL POSTSECONDARY STEM EDUCATION PROGRAMS ADDRESS WORKFORCE NEEDS TO SOME EXTENT

Most Federal Postsecondary STEM Education Programs Consider Workforce Needs, Including Jobs, Diversity, and Innovation

Eighty-eight percent[24] of the 124 federal postsecondary STEM education programs[25] that responded to our survey indicated that meeting one or more of the workforce needs we identified, such as promoting a diverse workforce, was a stated objective of the program.[26] An additional 11 percent of postsecondary programs indicated that meeting at least one workforce need was a potential benefit of their program activities, even if it was not a stated objective.[27] The most common stated objective was to prepare postsecondary students for a career in a STEM field. See figure 11 for fiscal year 2012 obligations associated with the various workforce needs.

Jobs

Eighty percent of the 124 federal postsecondary STEM programs that responded to our survey said that they focused on specific STEM occupations—41 percent as a stated objective and an additional 39 percent as a potential benefit of the program.[28] Almost three-quarters of obligations by grant-making programs with a stated objective to increase the numbers of workers in specific STEM occupations were made by programs that said they gave preference to applicants with the same goal.[29] Programs generally reported that they chose occupations according to market demand, their agency's mission, or both.

Fifty-six percent of the programs (25 percent of obligations) that focused on specific fields said that they chose occupations based on market demand.[30] Most of these programs reported that they identified high-demand occupations using national data and their own formal and informal research, such as networking with local industries. (See figure 12.) Some programs also indicated that they obtained information about high-demand occupations through partnerships with other organizations, such as industry groups that conduct national workforce needs assessments.

Along with high-demand occupations, most of the STEM education programs (85 percent of programs, 65 percent of obligations) that focused on specific fields reported that they chose occupations related to the agency's mission.[31] For example, the Department of Energy's mission corresponds to some specific STEM fields, such as energy science and nuclear physics, and the majority of programs from this agency said that they focus on mission-related occupations. Furthermore, one-third of the programs that target specific fields told us they focus solely on occupations related to their agency's mission instead of on high-demand occupations. One of the 13 programs we studied in depth—the National Institutes of Health's Ruth L. Kirschstein National Research Service Awards for Individual Predoctoral Fellows program—aims to address needs for biomedical, behavioral, and clinical research in the country. For this reason, grant guidance states that applicants must propose projects in research areas that fall under the agency's scientific mission. Additionally, 60 percent of postsecondary STEM education programs, representing 59 percent of obligations, said that they prepared students for jobs at their own agencies. While this may meet some workforce needs, the agency creates its own closed loop of trainees, job openings, and employees, and does not necessarily try to provide STEM workers to the broader workforce.

Diversity

In addition to preparing students for STEM jobs, we identified several other workforce needs that federal STEM education programs reported addressing. For example, experts and agency officials told us that programs that increase the diversity of the STEM workforce, prepare students for innovation and emerging fields, or provide STEM skills to students who do not obtain STEM degrees can contribute to American competitiveness in other ways. Experts also said that federal STEM programs are uniquely positioned to meet some of these broader workforce needs, which may not be provided by the marketplace alone.

A majority of the postsecondary STEM education programs in our survey indicated that they focus on increasing the numbers of minority, disadvantaged, or under-represented groups in the STEM workforce: 38 percent (45 percent of obligations) as a stated program objective, and 54 percent (48 percent of obligations) as a potential benefit of the program. Programs with a stated objective to increase the diversity of the STEM workforce most frequently reported that they served one or more under-represented racial or ethnic groups and people from economically disadvantaged backgrounds, and least frequently reported serving women.[32] Additionally, 77 percent of obligations by grant-making programs that responded to our survey were made by programs that reported that they gave preference to grant applicants that intend to increase the number of STEM workers from minority, disadvantaged, or under-represented groups.[33]

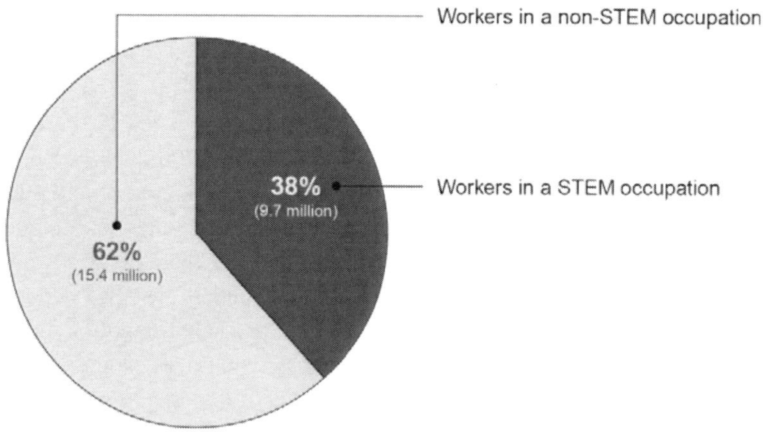

Source: GAO analysis of American Community Survey data.

Note: Estimates of the percentages have margins of error that are within plus or minus 0.2 percentage points. Population estimates shown in this figure have margins of error that are within plus or minus 0.01 percent of the estimate.

Figure 10. Percentage of Workers with Science, Technology, Engineering, and Mathematics (STEM) Bachelor's Degrees in STEM and Non-STEM Occupations, 2012.

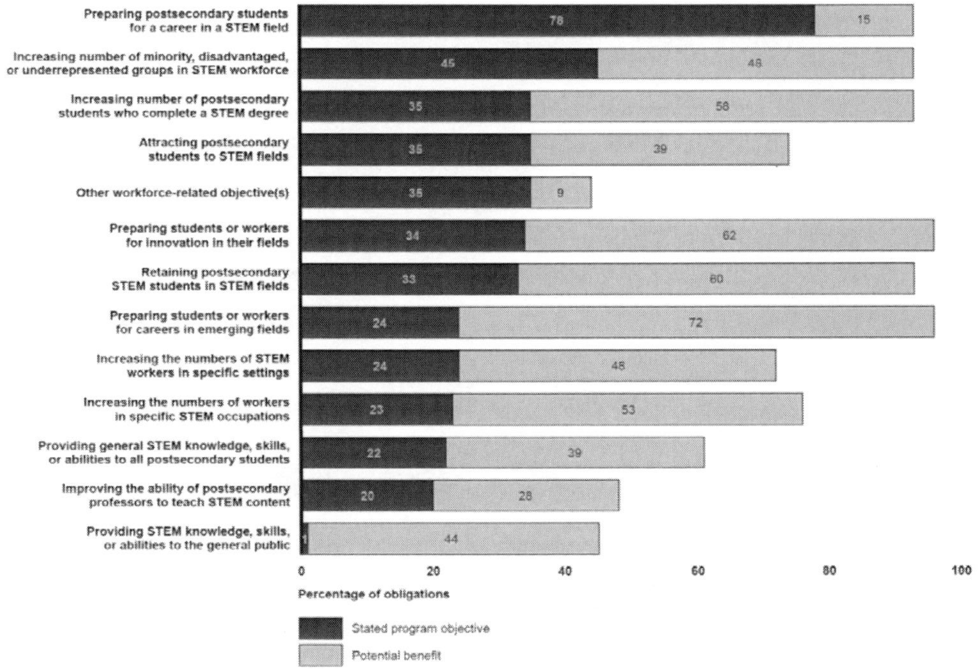

Source: GAO survey of STEM education programs.

Note: For this figure, N=124 programs. Programs may have more than one stated objective or potential benefit, so total obligations cannot be summed across the various workforce needs.

Figure 11. Percent of Reported Fiscal Year 2012 Federal Postsecondary Science, Technology, Engineering, and Mathematics (STEM) Education Obligations Dedicated to Workforce Needs.

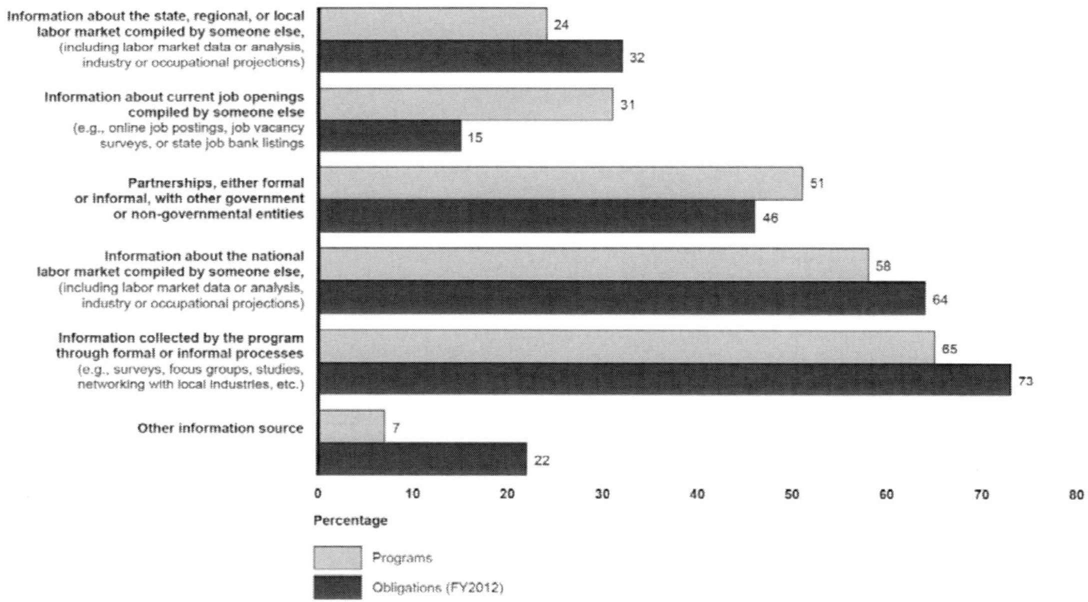

Source: GAO survey of STEM education programs.

Note: This figure includes only those 55 postsecondary STEM education programs (N=55 programs) that reported that they focused on high-demand occupations. Not included in this figure were 33 programs that focused only on occupations related to their missions, 11 programs that did not specify how they chose occupations, and 25 programs that indicated that they do not have a stated objective or potential benefit of increasing the number of workers in specific occupations.

Figure 12. Information Sources Used by Science, Technology, Engineering, and Mathematics (STEM) Education Programs that Target Occupations Based on High Market Demand, Fiscal Year 2012.

Four of the thirteen programs we studied in depth reported that they were primarily intended to serve minority, disadvantaged, or under-represented groups in STEM fields. For example, the Department of Education's Hispanic-Serving Institutions STEM and Articulation Programs award grants to postsecondary institutions with undergraduate student bodies that are at least 25 percent Hispanic. Grantees may create new coursework, improve infrastructure, develop research opportunities for students, or provide outreach and support services to students in order to encourage their pursuit of STEM degrees. Additionally, the National Science Foundation's Louis Stokes Alliances for Minority Participation program seeks to increase the numbers and qualifications of STEM graduates from under-represented groups. Grantees are allowed wide latitude to design projects that improve the undergraduate educational experiences of students and facilitate their transfer from 2-year to 4-year postsecondary institutions.

Innovation

Innovation is another workforce need that most federal postsecondary STEM programs reported that they aim to meet. In fact, among postsecondary STEM programs responding to our survey, preparing students or workers for innovation in their field and for careers in emerging STEM fields were the workforce needs with the highest reported obligations. However, although 95 percent of the 124 STEM programs that responded to our survey (97 percent of obligations) indicated that they intended to prepare people for innovation in their fields or for emerging STEM fields, 59 percent (61 percent of obligations) considered innovation to be a potential benefit rather than a stated objective. For example, the National Science Foundation and the National Institutes of Health both consider innovation in their agency-wide grant-making guidance. Additionally, the National Science Foundation sometimes creates agency-wide priorities for funding certain emerging fields, such as clean energy.

Programs Addressed Workforce Needs in Various Ways, and Some Measured Workforce-Related Outcomes

Federal postsecondary STEM education programs that responded to our survey indicated that they provided a range of services. The most common services they reported included research opportunities, internships, and mentorships. (See figure 13.)

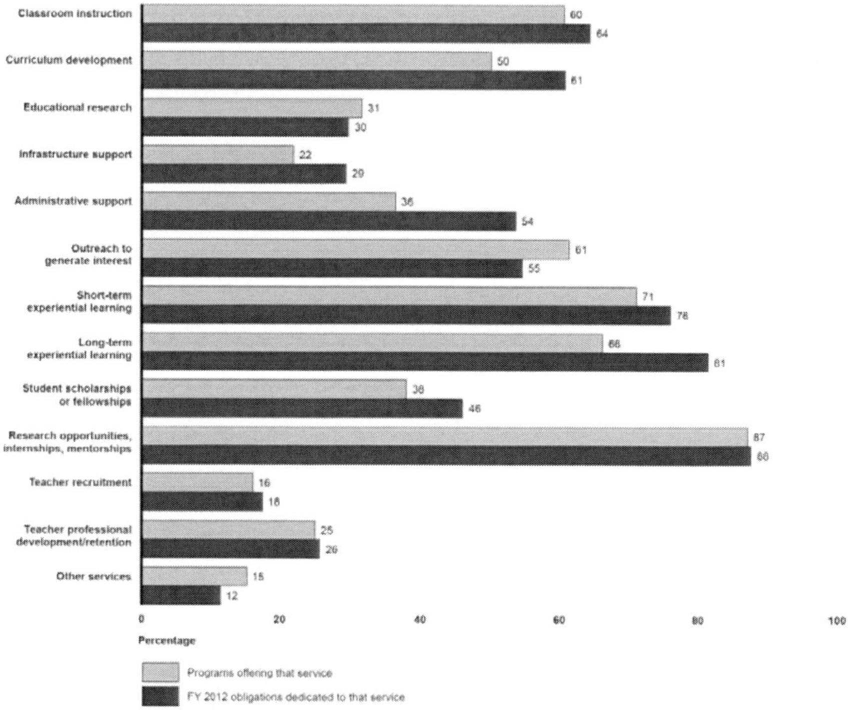

Source: GAO survey of STEM education programs.

Figure 13. Services Provided by Federal Postsecondary Science, Technology, Engineering, and Mathematics (STEM) Education Programs in Fiscal Year 2012.

Eighty percent of the 124 postsecondary STEM education programs that responded to our survey, representing 88 percent of obligations, said they tracked their success at meeting workforce needs using at least one outcome-based measure. Degree attainment, number of students pursuing STEM coursework, number of students taking a STEM job, and participant satisfaction were the most commonly reported outcomes. For example, the National Institutes of Health produced a report focused on the workforce outcomes of biomedical students, the majority of whom receive support from the National Institutes of Health at some point in their graduate careers.[34]

However, some programs did not measure an outcome or output that directly related to their stated objectives. For example, of the 78 postsecondary programs with a stated program objective to prepare students for STEM careers, 53 percent (45 percent of obligations) reported that they did not track the number of their students who took a job in a STEM field. Similarly, of the

49 programs with a stated program objective to increase the numbers of STEM graduates, 39 percent (43 percent of obligations) reported that they did not measure the educational attainment of their program participants. These data are consistent with our 2012 STEM report, in which we found that STEM education programs' outcome measures were not clearly reflected in the performance planning documents of most agencies.[35] As we recommended in 2012, the National Science and Technology Council recently issued guidance to help agencies better incorporate their STEM education efforts and the goals from the government-wide STEM Education 5-Year Strategic Plan into their agency performance plans and reports. As agencies follow the guidance, improve their outcome measures, and focus on the effectiveness of the programs, more programs may measure outcomes directly related to their stated program objectives, such as preparing students for STEM careers.

MOST FEDERAL K-12 STEM EDUCATION PROGRAMS INCLUDE ACTIVITIES TO PREPARE STUDENTS FOR FUTURE STEM EDUCATION

According to our survey, preparing students for postsecondary education in a STEM field is either a stated program objective or a potential secondary benefit of almost all federal K-12 STEM education programs.[36] Specifically, out of 30 federal K-12 STEM education program respondents to our survey, 13 programs (50 percent of K-12 program obligations)[37] reported that preparing students for postsecondary STEM education is a stated program objective, while 16 programs reported that it is a potential benefit of the program.[38]

Of the six federal K-12 STEM education programs we selected to review in more detail, four programs—Advanced Technological Education, Discovery Research K-12, Math and Science Partnership, and Upward Bound Math-Science—reported that preparing students for postsecondary STEM education is a stated objective of the program. Upward Bound Math-Science programs, which are based in institutions of higher education, work closely with students to strengthen their math and science skills in order to prepare and encourage them to pursue postsecondary degrees in math and science.[39] According to an official from an Upward Bound Math-Science program we visited in California, the program is not specifically intended to prepare students for the STEM workforce, but it emphasizes helping students understand the varied career opportunities available to them in math and science fields. Officials from another Upward Bound Math-Science program we visited said they try to connect their students with practitioners in the field, since it is important for students to have role models in STEM occupations who hail from similar backgrounds.

In our survey, 18 of the 30 federal K-12 STEM education programs (approximately 77 percent of K-12 program obligations) reported that improving the ability of K-12 teachers to teach STEM content is a stated program objective.[40] Several experts have noted that one challenge at the K-12 level is that STEM teachers sometimes do not have sufficient content knowledge to effectively teach these subjects, and that the federal government can play an important role by supporting professional development for STEM teachers and encouraging more college graduates in STEM fields to pursue teaching careers.

Four of the federal K-12 STEM education programs we reviewed in detail— Advanced Technological Education, Discovery Research K-12, Math and Science Partnership, and the Mathematics and Science Partnerships program[41]—reported that improving the ability of K-12 teachers to teach STEM content is a stated program objective. The Mathematics and Science Partnerships program provides formula grants to states, which in turn award competitive grants to partnerships that enhance the content knowledge and teaching skills of math and science teachers. A Mathematics and Science Partnerships grantee we met with in Texas established regional networks across the state in which mentor teachers provided professional development and mentoring to participating teachers. Similarly, the Discovery Research K-12 program supports research projects that address a need in STEM education at the pre-kindergarten through 12th grade levels, particularly programs that explore unconventional approaches to teaching and learning. Researchers we met with were exploring how computational models could be used to make decisions about resource allocation to optimize learning in STEM classes. For example, the model might be used to calculate optimal student-teacher ratios given other factors, such as grade level, subject, and class composition.

In our survey, 7 of the 30 federal K-12 STEM education programs (approximately 26 percent of K-12 program obligations) reported that providing students with STEM knowledge, skills, and abilities, without the explicit goal of preparing them for postsecondary STEM education or a STEM career, is a stated program objective. According to recent research, exposing students to STEM content and encouraging their interest in STEM disciplines at an early age is important in order to increase the likelihood that they remain engaged with STEM later in life. The National Science Foundation's Advancing Informal STEM Learning program provides grants to organizations working on innovative projects intended to expose students to STEM content outside the classroom. A museum we visited in California received an Advancing Informal STEM Learning grant to develop an outdoor bilingual science exhibit and related curriculum targeted towards Latino students in the San Francisco area. Officials told us the exhibit is geared towards students who may not generally visit the museum.

Federal K-12 STEM education programs provide a variety of educational services in order to achieve their objectives. The services identified most often in our survey included classroom instruction; curriculum development; outreach to generate student interest in STEM fields; short-term experiential learning activities; and teacher professional development or retention activities (see figure 14).

In our survey, 25 of the 30 federal K-12 STEM education programs (approximately 89 percent of K-12 program obligations) reported that they tracked or monitored program outcome measures in 2012. However, as with the federal postsecondary STEM education programs some K-12 programs are not measuring outcomes directly related to their stated objectives. For example, of the 13 K-12 programs that reported having a stated program objective to prepare students for postsecondary STEM education, 10 programs said they did not track student educational attainment or the number of students who pursued coursework in STEM fields. Of the 18 programs that reported that improving the ability of K-12 teachers to teach STEM content was a stated program objective, 6 programs said they did not monitor teacher improvement and performance in STEM education instruction or the number of qualified teachers teaching STEM.

K-12 STEM education program grantees we met with monitored some programmatic outcomes. For example, an official from an Upward Bound Math-Science program we visited told us that each program is required to submit an annual report to Education, including data on performance outcomes such as the number of participants who graduate from high school, pursue postsecondary degrees in STEM fields, and graduate from college within 6 years.[42] The official said that all but one of the participants in the program's first cohort graduated from high school and enrolled in college. Further, officials from a Mathematics and Science Partnerships grantee told us that—in addition to mandatory reporting to Education on performance outcomes, such as the number of teachers trained through the program and the extent to which teachers' test scores showed statistically significant gains—they were implementing an initiative to correlate programmatic data with student outcomes across the state, as measured by teacher self-reporting and statewide assessments. The initial phase of the analysis, based on teacher self-reporting, found that the students whose teachers had participated in the program outperformed their peers in several STEM subjects. In addition, officials from the museum exhibit in California funded by the Advancing Informal STEM Learning program said assessments were planned for every stage of the project, including a summative evaluation to review the extent to which it may have influenced Latino youth awareness of and engagement with STEM content. Officials said the evaluation would be completed in January of 2015.

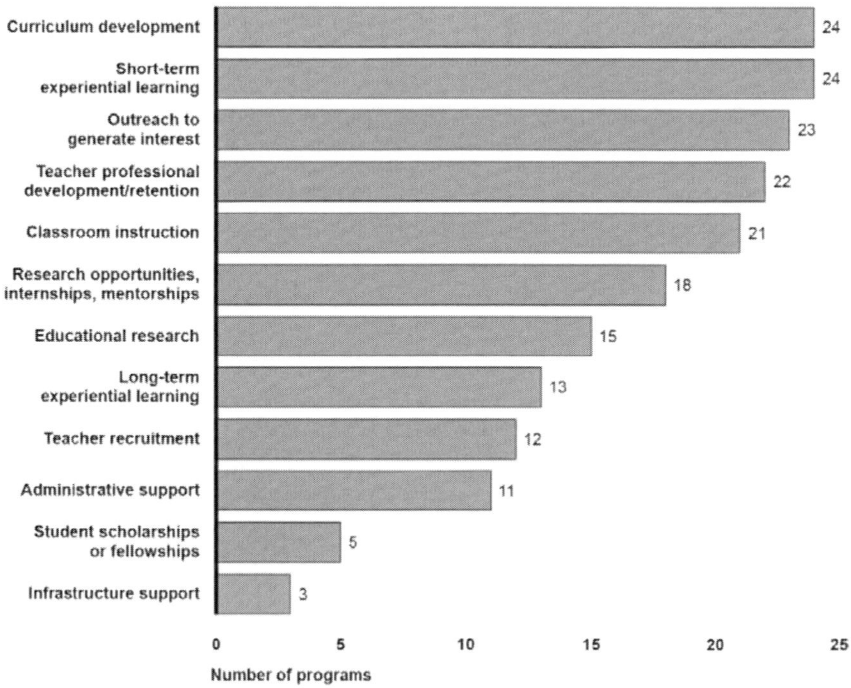

Source: GAO survey of STEM education programs.

Figure 14. Number of Kindergarten-12th grade Science, Technology, Engineering, and Mathematics (STEM) Education Programs Providing Various Services in Fiscal Year 2012.

CONCLUDING OBSERVATIONS

It is difficult to determine whether there has been a shortage or a sufficient supply of STEM workers in the United States and, consequently, to define the appropriate role the federal government should play in increasing the number of STEM-educated workers. There is not a one-to-one match between STEM graduates in a specific field and corresponding STEM jobs because not all people with STEM degrees pursue careers in their fields of study, whether by choice or because of limited employment opportunities in the field. Regardless of career choices, the rigor of a STEM education may help promote a workforce with transferable skills and the potential to fuel innovation and economic growth. Federal postsecondary STEM education programs may help develop a workforce that will address issues that affect the population as a whole, such as researching diseases or improving defense capabilities. Additionally, federal K-12 STEM education programs may generate interest in STEM fields early in life, which could usher more students into the STEM pipeline and increase the likelihood that they will pursue STEM education and careers.

Although the administration has taken steps to consolidate and coordinate STEM education programs, numerous programs—spread across 13 agencies—remain. As the administration continues to consolidate and eliminate STEM education programs, it risks making decisions without considering the efficacy of these programs because many federal STEM education programs are not measuring their outcomes. However, the guidance recently issued by the National Science and Technology Council could help agencies better incorporate their STEM education efforts and the goals from the government-wide 5-year STEM strategic plan into their agency performance plans and reports. This will enable agencies to better assess which STEM education efforts are successful in contributing to agency-wide performance goals and supporting the overall federal STEM effort.

AGENCY COMMENTS AND OUR EVALUATION

We provided a draft of this product for comment to the Departments of Defense, Education, Energy, and Health and Human Services; National Science Foundation; and Office of Management and Budget. All provided technical comments except the Department of Defense, which indicated that it had no comments. We incorporated the technical comments as appropriate.

Melissa Emrey-Arras
Director, Education, Workforce, and Income Security Issues

APPENDIX I. OBJECTIVES, SCOPE, AND METHODOLOGY

Our research objectives were to review (1) recent trends in the number of degrees and jobs in Science, Technology, Engineering, and Mathematics (STEM) fields, (2) the extent to which federal postsecondary STEM education programs take workforce needs into consideration, and (3) the extent to which federal kindergarten-12th grade (K-12) STEM education programs prepare students for postsecondary STEM education.

To inform all of our objectives, we reviewed relevant federal laws and regulations. We also reviewed relevant literature and past reports on STEM education, including our 2012 STEM report[43] and the National Science and Technology Council's Strategic Plan for federal STEM education programs. In addition, we interviewed STEM experts and officials from the Office of Science and Technology Policy and several other federal agencies that administer STEM education programs to gather information on their STEM education efforts. We attended a STEM education conference to gather additional perspectives about federal STEM education programs.

Analysis of Education and Workforce Data

Analysis of the Integrated Postsecondary Education Data System (IPEDS) Data

To examine recent trends in the number of STEM degrees awarded, we analyzed data from IPEDS from the Department of Education's National Center for Education Statistics. IPEDS is a system of interrelated surveys conducted annually to gather information from every college, university, and technical and vocational institution that participates in federal student financial aid programs. The Higher Education Act of 1965, as amended, requires institutions of higher education that participate in federal student aid programs to complete IPEDS surveys.[44] IPEDS provides institution-level data in such areas as enrollment, program completions, faculty, staff, and finances.

Specifically, we analyzed 10 years of data from the IPEDS program completions component, from the July 2002-June 2003 academic year to the July 2011-June 2012 academic year. The program completions component provides data on the number of degrees awarded by each institution for each program of study. We analyzed the data to determine the number of degrees awarded nationally in STEM and non-STEM programs of study in this time period, the number awarded in our three STEM categories, and the number awarded in selected STEM fields. We included degrees awarded for both first and second majors in our analysis. Our results represent the number of degrees awarded, not the number of individuals who obtained degrees. We assessed the reliability of the IPEDS data we used by reviewing relevant documents and past GAO reviews of the data and conducting electronic testing. On the basis of this assessment, we concluded that the data were sufficiently reliable for our reporting purposes.

In conducting our analysis, we classified each program of study in the IPEDS data as STEM or non-STEM. We used as guidance work conducted by the Census Bureau to classify fields of study as science and engineering or science- and engineer-related in the American Community Survey (ACS) data. This helped to ensure that we were consistent with the fields we defined as STEM in both our IPEDS and ACS analyses. We further classified these STEM fields into our three STEM categories of Core STEM, Healthcare STEM, and Other STEM. See table 1 below for the fields of study we classified as STEM and how we classified them into our three STEM categories. We also aggregated detailed programs of study into broader STEM fields, generally based on the first two digits of the Classification of Instructional Programs code (the classification system that IPEDS uses to define programs of study). For example, Classification of Instructional Programs codes beginning with 11 represent programs of study under the category of “computer and information sciences and support services.” The information we present on numbers of computer science/information technology (IT) degrees comes from aggregating the number of degrees awarded for Classification of Instructional Programs codes that begin with 11. For life sciences, mathematics and statistics, and social sciences, we combined programs of study from multiple 2-digit Classification of Instructional Programs code categories (see table 4 for the fields we combined).

Table 4. Fields of Study We Classified as Science, Technology, Engineering, and Mathematics (STEM)

STEM Category	Programs of Study, based on the Classification of Instructional Programs 2010
Core STEM	Life sciences (includes biological and biomedical sciences; agricultural sciences, i.e., animal sciences, food science and technology plant sciences, and soil sciences; natural resources and conservation, i.e., environmental studies/science, fisheries science and management, forestry, wildlife and wildlands science and management)
	Computer and information sciences and support services
	Engineering
	Technician (includes engineering technologies and engineering-related fields, science technologies/technicians)
	Mathematics and statistics (including actuarial science)
	Physical sciences
Healthcare STEM	Health professions and related programs
Other STEM	Social sciences (includes natural resource economics and agricultural economics; area, ethnic, cultural, gender, and group studies; linguistics; public policy analysis; history and philosophy of science and technology; social sciences).
	Architecture and related services
	STEM teacher education
	Multi-/interdisciplinary studies (i.e., systems science, nutrition sciences, behavioral sciences, gerontology, international/global studies)
	Psychology

Source: GAO analysis of the Integrated Postsecondary Education Data System

Note: We did not include construction trades, mechanic and repair technologies/technicians, or precision production in our definition of STEM fields because they were not defined as STEM in the information sources that informed our categorization. We also did not include industry certifications or licenses. We note that Congress recently changed the name of the Office of Vocational and Adult Education to the Office of Career, Technical, and Adult Education.

Analysis of Occupational Employment Statistics (OES) Data

To examine trends in STEM occupations, we analyzed the Bureau of Labor Statistics' OES data from the May 2004 survey to the May 2012 survey. The OES program surveys establishments to produce estimates of employment and wages for specific occupations.[45] We began our analysis with the May 2004 data because that was the first year that all occupations in the OES were classified based on the Standard Occupational Classification (SOC) system. We conducted our analysis to identify trends in the number of jobs and the average wages in STEM and non-STEM occupations from 2004 to 2012. We assessed the reliability of the OES data by reviewing relevant documents, interviewing Bureau of Labor Statistics officials, and conducting electronic testing of the data. Based on our assessment, we concluded that the OES data were sufficiently reliable for our reporting purposes.

We classified occupations as STEM and non-STEM based on the SOC Policy Committee's *Options for Defining STEM (Science, Technology, Engineering, and Mathematics) Occupations Under the 2010 Standard Occupational Classification System.*[46] This document sets out several options for defining STEM occupations. Any occupation that was included in any of the SOC Policy Committee's options was classified as STEM in our analysis. All other occupations were classified as non-STEM. We also classified occupations into our three STEM categories of Core STEM, Healthcare STEM, and Other STEM based some of the options presented by the SOC Policy Committee. Specifically:

- Occupations categorized by the SOC Policy Committee as "Life and Physical Science, Engineering, Mathematics, and Information Technology Occupations" were classified as Core STEM occupations in our analysis. These include postsecondary teachers, managers, technicians, and scientists in these fields, as well as sales representatives for technical and scientific products and sales engineers.[47]
- Occupations categorized by the SOC Policy Committee as "Health Occupations" were classified as Healthcare STEM occupations in our analysis. These included health diagnosing and treating practitioners, health technologists and technicians, postsecondary health teachers, and medical and health services managers. It does not include healthcare support occupations (e.g., health aides, nursing assistants).
- Occupations categorized by the SOC Policy Committee as "Social Science Occupations and "Architecture Occupations" were classified as Other STEM occupations in our analysis. These include scientists and researchers, architects and related professions, assistants, and postsecondary teachers in these fields.

The SOC Policy Committee's *Options for Defining STEM Occupations* was based on occupations defined under the 2010 SOC, while the 2004 to 2009 OES data used a slightly different occupational classification system (the 2000 SOC). We used Bureau of Labor Statistics crosswalks between the 2000 SOC and the 2010 SOC to identify the appropriate STEM occupations throughout the period of our study.[48] We also combined detailed occupations into broader occupational groups based on the first two or three digits of the SOC codes and presented employment and wage trends for these occupational groups (e.g., computer/IT occupations). Specifically, our categories of STEM management and STEM sales in figure 6 of our report combine occupations under the two digit-SOC codes 11 (management occupations) and 41 (sales and related occupations). Other occupational categories presented in figure 6 combine occupations based on the first three digits of the SOC codes (e.g., our computer/IT category combines occupations beginning with SOC code 15-1, computer occupations).

To minimize respondent burden, the OES survey is conducted on a 3-year cycle that ensures that most establishments are surveyed at most once every three years. OES estimates are produced annually, but each year's estimates are based on surveys conducted over a 3-year period. Following Bureau of Labor Statistics guidance for using OES data that are at least two or three years apart when examining trends over time, we present results for alternate years in appendix III (for May 2004, 2006, 2008, 2010, and 2012). We calculated standard errors for our estimates based on the relative standard errors that the Bureau of Labor Statistics provided for each employment and mean wage estimate for each occupation.

Analysis of ACS Data

We analyzed the data from the Census Bureau's ACS to examine the unemployment rates of those in STEM and non-STEM occupations, as well as the educational backgrounds at the bachelor's degree level of those in STEM and non-STEM occupations. The ACS is an ongoing national survey which replaced the decennial census long-form questionnaire as a source for social, economic, demographic, and housing information. About 3 million households are selected for the ACS each year. The ACS questionnaire asks about the kind of work people in the household were doing in their most recent job if they worked in the last 5 years (i.e., their occupation). It also asks about the highest degree or level of school a person has completed. If the person has completed a bachelor's degree or higher, the ACS asks for the specific major(s) of any bachelor degree(s) the person has completed. The ACS also contains questions to produce estimates of the number of people who are employed, unemployed, and not in the labor force. We specifically analyzed data from the 1-year Public Use Microdata Samples for 2009 to 2012. We assessed the reliability of the data by reviewing relevant documentation and conducting electronic testing of the data. Based on our assessment, we concluded that the ACS data were sufficiently reliable for our reporting purposes.

The Census Bureau has its own system for coding occupations and fields of study in the ACS data, which are based on the SOC and the CIP, respectively. Census has also classified occupations as STEM and STEM-related (healthcare and architecture) and fields of study as science and engineering and science- and engineering-related. The Census Bureau's classifications of occupations are based on the SOC Policy Committee's *Options for Defining STEM Occupations*, though agency officials made some modifications due to their use of different coding systems. We considered any occupation that Census classified as STEM or STEM-related as STEM in our analysis of occupations, and any field of study they identified as science and engineering and science- and engineering-related as STEM in our analysis of degrees. As with our analysis of OES data, we classified occupations into our three STEM categories of Core STEM, Healthcare STEM, and Other STEM. We also combined detailed occupations and fields of study into broader categories. For example, we combined 11 specific occupations into our category of computer/IT occupations, and 6 different fields of study for the computer/IT major at the bachelor's degree level.

With regard to the unemployment rates we present, most of our estimates are for the civilian population in the labor force ages 16 and older. Our estimates of the educational background of those in STEM and non-STEM occupations are based on the population ages 22 and older. Our estimates of the unemployment rates of those in STEM and non-STEM occupations by educational background (in figure 6 of appendix III) are for the civilian population in the labor force ages 22 and older. The Bureau of Labor Statistics has found that ACS estimates of the unemployment rate can differ from estimates produced by the Current Population Survey, a monthly survey of about 60,000 households that is the nation's source of official government statistics on employment and unemployment. The Bureau of Labor Statistics states that a number of factors may account for the differences, including overall questionnaire differences, differing requirements in the two surveys with regard to whether an individual is actively looking for work, and differences in reference periods, modes of collection, and population controls.

We calculated standard errors for our estimates using the replicate weight method. For some estimates of the unemployment rate for specific occupational categories, the margin of error exceeded 30 percent of the estimates. We note these instances in our report.

Regression Analysis

In order to compare the wages and unemployment rates of workers in STEM and non-STEM occupations with comparable personal characteristics, we ran a series of wage regressions and unemployment regressions in which we controlled for human capital characteristics (age and education) and demographic characteristics (race, ethnicity, gender, citizenship, and veterans status) as well as the worker's broad occupational category.

We used the ACS for our wage and unemployment regression analyses. We restricted our analysis to full-time, full-year workers. We restricted our analysis to full-time workers because the ACS does not collect data on whether people are salaried or hourly workers, making it difficult to use the "usual weekly hours" variable. We restricted our analysis to full-year workers because the ACS also does not collect data on weekly wages, but on earnings from wages or salary in the past year. Not all people work a full year, and people who have been unemployed for part of the year will have annual earnings that do not reflect their annual salary or hourly rate of pay. When constructing our dependent variable, we took the natural log of annual wages.

For the unemployment regressions, the outcome variable is current labor force status. People who are currently unemployed are defined as unemployed; people who are currently working or on paid leave from work are defined as not unemployed; and people who are not in the labor force are excluded from the universe. The universe is also restricted to people ages 16-64, and excludes people who have no work experience or have not worked in the past 5 years because the ACS does not collect occupation for these people.

Both sets of regressions use linear models and the same set of covariates.

Definition of STEM Education Program

For the purposes of our study, we applied the definition of a federally-funded STEM education program used in previous GAO work.[49] Specifically, we defined it as a program funded in fiscal year 2012 by allocation or congressional appropriation that had not been subsequently terminated and included one or more of the following as a primary objective:

- attract or prepare students to pursue classes or coursework in STEM areas through formal or informal education activities (informal education programs provide support for activities provided by a variety of organizations that offer students learning opportunities outside of formal schooling through contests, science fairs, summer programs, and other means; outreach programs targeted to the general public were not included),
- attract students to pursue degrees (2-year, 4-year, graduate, or doctoral degrees) in STEM fields through formal or informal education activities,
- provide training opportunities for undergraduate or graduate students in STEM fields (this could include grants, fellowships, internships, and traineeships that are targeted to students; general research grants that are targeted to researchers that may hire a student to work in the lab were not considered a STEM education program),
- attract graduates to pursue careers in STEM fields,
- improve teacher (pre-service or in-service) education in STEM areas,
- improve or expand the capacity of K-12 schools or postsecondary institutions to promote or foster education in STEM fields, or
- conduct research to enhance the quality of STEM education programs provided to students.

In addition, STEM education programs may provide grants, fellowships, internships, and traineeships. While programs designed to retain current employees in STEM fields were not included, programs that fund retraining of workers to pursue a degree in a STEM field were included because these programs help increase the number of students and professionals in STEM fields by helping retrain non-STEM workers to work in STEM fields.

For the purposes of this study, we defined an organized set of activities as a single program even when its funds were allocated to other programs as well. Several programs had been eliminated or consolidated into new programs since our last inventory. We included programs that had been consolidated, but we did not include programs that had since been terminated. For a list of STEM education programs by agency, including consolidated programs, see appendix IV.

Program Selection

To identify federally-funded STEM education programs, first we developed a combined list of programs based on the findings of two previous STEM education inventories—one that we issued in 2012 and another completed by the National Science and Technology Council in 2011. Second, we shared our list with agency officials, along with our definition of STEM education program, and asked officials to make an initial determination about which programs should remain on the list and which programs should be added to the list. If agency officials indicated they wanted to remove a program from our list, we asked for additional information. For example, programs on our initial list may have been terminated or consolidated, or did not receive federal funds in fiscal year 2012.

Table 5. Science, Technology, Engineering, and Mathematics (STEM) Education Programs Reviewed In Depth

Department of Education	
Postsecondary programs	Hispanic Serving Institutions STEM and Articulation Programs
K-12 programs	Upward Bound Math-Science
	Mathematics and Science Partnerships
Department of Health and Human Services (National Institutes of Health)	
Postsecondary programs	Ruth L. Kirschstein National Research Service Award Institutional Research Training Grants
	Ruth L. Kirschstein NRSA for Individual Predoctoral Fellows, including Under-representedRacial/Ethnic Groups, Students from Disadvantaged Backgrounds, and Predoctoral Students with Disabilities
National Science Foundation	
Postsecondary programs	Graduate Research Fellowship Program
	Integrative Graduate Education and Research Traineeship Program
	Louis Stokes Alliances for Minority Participation
	Research Experiences for Undergraduates
K-12 programs	Advancing Technological Education
	Discovery Research K-12
	Advanced Informal Science Learning
	Math and Science Partnership

Source: GAO.

We reviewed additional information on the programs that were not included in our 2012 inventory of STEM education programs, mainly through agency websites, program materials, or discussions with program officials. On the basis of this additional information, we excluded programs that we found did not meet our definition of a STEM education program. We also included screening questions in the survey to provide additional verification that the programs met our definition of a STEM education program. Of the 170 programs on our original survey distribution list, seven programs did not pass our screening questions because they had been eliminated since 2012, and we determined that another five did not meet our definition of a STEM education program.[50] In total, we identified 158 federal STEM education programs.

To provide more details about some of the STEM education programs with the highest reported obligations, we conducted a more in-depth review of 13 of the largest STEM education programs from three agencies: the National Science Foundation, the Department of Education, and the National Institutes of Health at the Department of Health and Human Services.[51] Seven of the selected programs served postsecondary students or institutions and six programs served K-12 students or teachers (see table 5). We reviewed documentation from each program, interviewed agency officials, and conducted site visits with grantees in Austin and San Francisco and phone interviews with grantees in Boston. We chose these sites based on geographic diversity and the prevalence of federal STEM grantees.

Survey

Design and Implementation

We developed a web-based survey to collect information on federal STEM education programs. The survey included questions on program objectives, occupations targeted, methods used to identify targeted occupations, and factors considered when selecting grantees. We created a list of possible workforce needs using input from experts, program officials, and grantees, and asked federal STEM education programs to indicate whether each possible workforce need was a stated program objective, a potential benefit of the program, or neither. The survey also asked programs to update information provided in our survey for the 2012 report on target groups served, services provided, outcome measures, and obligations. To minimize errors arising from differences in how questions might be interpreted and to reduce variability in responses that should be qualitatively the same, we conducted pretests with six different programs in August and September 2013. To ensure that we obtained a variety of perspectives on our survey, we selected programs from six different agencies that differed in program scope, objectives, services provided, and target groups served. An independent GAO reviewer also reviewed a draft of the survey prior to its administration. On the basis of feedback from these pretests and independent review, we revised the survey in order to improve its clarity.

After completing the pretests, we administered the survey. On October 29 or November 13, 2013, we sent an e-mail message to the officials responsible for the 158 programs selected for our review that informed them that the survey was available online. In that e-mail message, we also provided them with unique passwords and usernames. We made telephone calls to officials and sent them follow-up e-mail messages, as necessary, to clarify their responses or obtain additional information. We received completed surveys from 154 programs, for a 97 percent response rate.[52] We collected survey responses through February 14, 2014.

Of the 154 federal STEM education programs that responded to our survey, 124 programs in 13 agencies primarily served students and teachers at the postsecondary level.[53] According to our survey, these programs' reported fiscal year 2012 obligations ranged from zero to $348 million and totaled $1.9 billion. We identified 30 programs in 10 agencies that primarily serve students and teachers at the K-12 level.[54] According to our survey, these programs reported obligations totaling approximately $685 million in fiscal year 2012 in amounts ranging from $1,200 to $148 million.

Analysis of Responses and Data Quality

We used standard descriptive statistics to analyze survey responses. The STEM education programs in our survey received widely varying amounts of federal funding. This introduced the possibility that a few very large programs—accounting for the majority of obligations—could pursue one activity, while many small programs—accounting for the majority of programs but a small proportion of obligations—could pursue another activity. To accurately capture the survey data, we analyzed it both in terms of the percentage of programs answering each question and the corresponding percentage of obligations. In cases where these proportions differed, we presented both. Amounts obligated for each program for fiscal year 2012 were reported to us by agency officials in response to our survey. We did not independently verify this information.

Because this was not a sample survey, there are no sampling errors. To minimize other types of errors—commonly referred to as nonsampling errors—and to enhance data quality, we employed recognized survey design practices in the development of the survey and in the collection, processing, and analysis of the survey data. For instance, as previously mentioned, we pretested the survey with federal officials to minimize errors arising from differences in how questions might be interpreted and to reduce variability in responses that should be qualitatively the same. We further reviewed the survey to ensure the ordering of survey sections was appropriate and that the questions within each section were clearly stated and easy to comprehend. To reduce nonresponse, another source of nonsampling error, we sent out e-mail reminder messages to encourage officials to complete the survey. To assess the reliability of data provided in our survey, we performed automated checks to identify inappropriate answers. We further reviewed the data for missing or ambiguous responses and followed up with agency officials when necessary to clarify their responses. While we did not verify all responses, on the basis of our application of recognized survey design practices and follow-up procedures, we determined that the data used in this report were of sufficient quality for our purposes.

Appendix II. Demographics of Science, Technology, Engineering, and Mathematics (STEM) Degree Recipients

The figures below show demographic information for students who received STEM degrees in the 2011-2012 academic year.

STEM Degrees Awarded to Non-Resident Alien Students

Overall, degrees awarded to non-resident alien students—students in the United States on temporary visas—comprised 5 percent of all STEM degrees and 4 percent of all non-STEM degrees awarded in the 2011-2012 academic year (see figure 15). However, degrees awarded to non-resident alien students represented a larger share of Core STEM degrees (11 percent) and a smaller share of Healthcare degrees (1 percent).

Non-resident alien students were particularly concentrated at the graduate degree levels in Core STEM fields, receiving 36 percent of master's degrees awarded and 42 percent of doctorate or professional degrees in Core STEM fields in the 2011-2012 academic year (see figure 16).

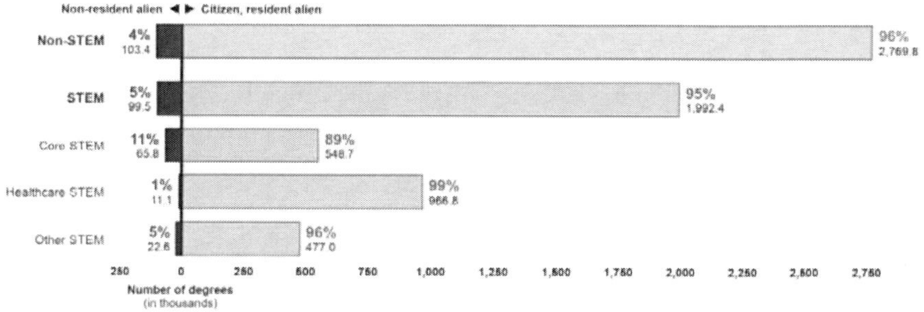

Source: GAO analysis of data from the Integrated Postsecondary Education Data System.

Figure 15. Degrees Awarded to Non-Resident Alien Students and to Citizens and Resident Aliens, 2011-2012 Academic Year.

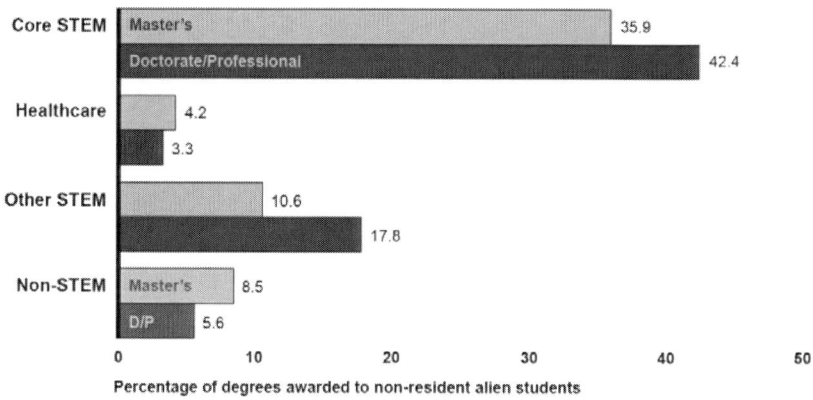

Source: GAO analysis of Integrated Postsecondary Education Data System (IPEDS) data.

Figure 16. Percent of Master's and Doctorate/Professional Degrees Awarded to Non-Resident Alien Students, 2011-2012 Academic Year.

Table 6. Science, Technology, Engineering, and Mathematics (STEM) Fields of Study in which Non-Resident Alien Students Received Over 30 Percent of the Degrees Awarded in the 2011-2012 Academic Year

STEM Field	Award Level	Degrees Awarded to Non-Resident Alien Students	Degrees Awarded to Citizens and Resident Aliens	Percentage of Degrees Awarded to Non-Resident Alien Students
Core STEM Fields				
Computer science/IT	Master's	9,402	12,068	44
	Doctorate	873	826	51
Engineering	Master's	17,192	24.316	41
	Doctorate	4,892	3,877	56
Mathematics	Master's	2,712	3,935	41
	Doctorate	820	854	49
Physical sciences	Doctorate	2,155	3,244	40
Other STEM Fields				
Architecture	Doctorate	102	106	49
Social sciences	Doctorate	1,526	3,271	32

Source: GAO analysis of data from the Integrated Postsecondary Education Data System.

Table 6 lists the STEM fields of study and degree levels in which non-resident alien students comprised more than 30 percent of the degrees awarded.

STEM Degrees by Gender

Overall, most (63 percent) of the STEM degrees awarded in the 2011-2012 academic year were awarded to women. However, as figure 17 shows, while women received the large majority (82 percent) of Healthcare STEM degrees that year, men received the majority of Core STEM degrees (68 percent).[55]

Among the Core STEM fields, men received the majority of degrees in computer science/information technology, engineering, technician, mathematics, and physical science fields.[56] Women received the majority of life sciences degrees (see figure 18).

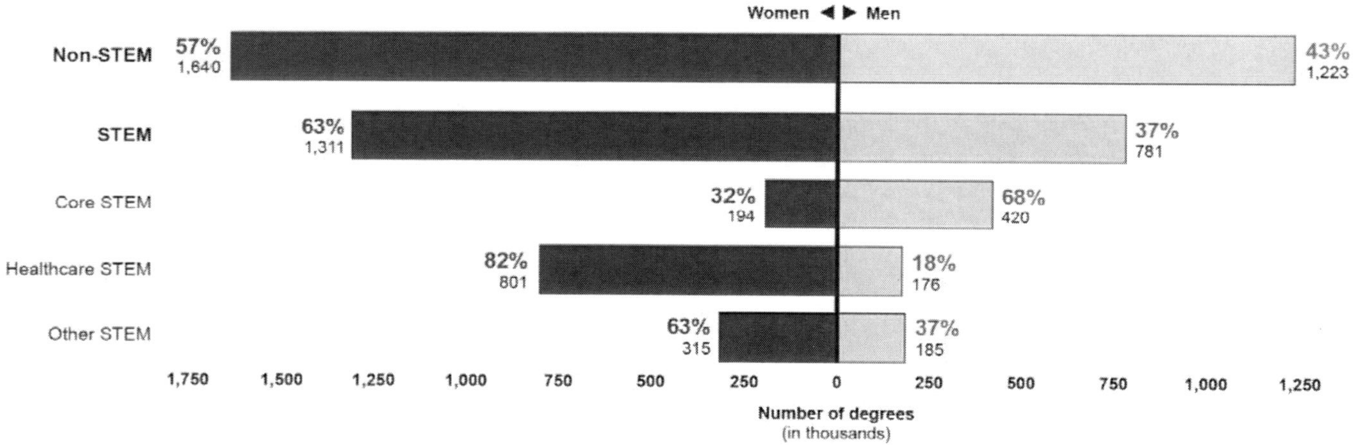

Source: GAO analysis of data from the Integrated Postsecondary Education Data System.

Figure 17. Science, Technology, Engineering, and Mathematics (STEM) and Non-STEM Degrees Awarded, by Gender, 2011-2012 Academic Year.

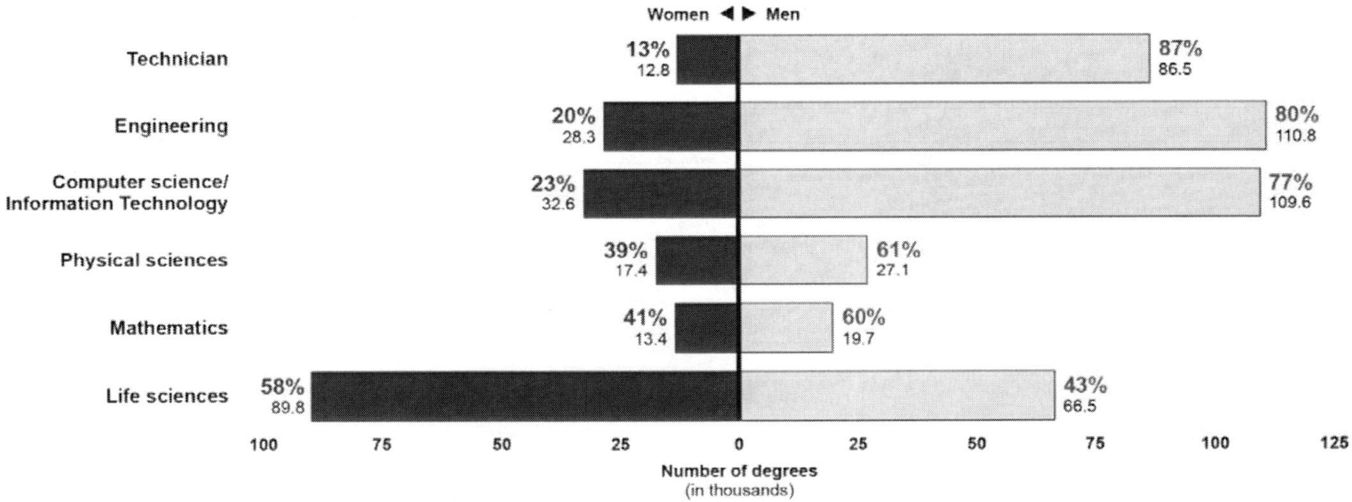

Source: GAO analysis of data from the Integrated Postsecondary Education Data System.

Note: Percentages may not total 100 due to rounding.

Figure 18. Degrees Awarded in Core Science, Technology, Engineering, and Mathematics (STEM) Fields, by Gender, 2011-2012 Academic Year.

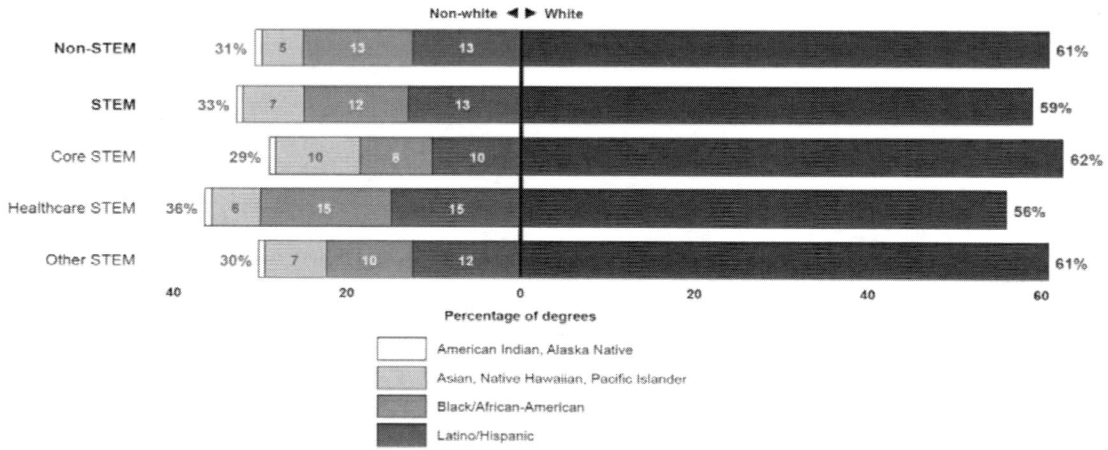

Source: GAO analysis of data from the Integrated Postsecondary Education Data System.

Note: This figure presents information on degrees awarded to citizens and resident aliens only. It does not include degrees awarded to non-resident aliens. Percentages for the non-white demographic groups may not total the percentages on the left of each bar due to rounding. The percentages for degrees awarded to American Indian, Alaska Native students are: 0.8 percent for non-STEM, 0.7 percent for STEM, 0.7 percent for Other STEM and Core STEM, and 0.8 percent for Healthcare. Not shown in the figure are degrees awarded to citizens and resident aliens of more than one race or of unknown race. As a result, the percentages on the left of each bar and the percentages on the right of each bar do not total 100. In the 2011-2012 academic year, degrees received by students of more than one race comprised: 1.3 percent of STEM degrees awarded to citizens and residents, 1.4 percent of non-STEM degrees, 1.4 percent of Core STEM degrees, 1.3 percent of Healthcare degrees, and 1.8 percent of Other STEM degrees. Degrees received by students whose race is unknown comprised 6.9 percent of STEM degrees awarded to citizens and resident aliens, 7.3 percent of non-STEM degrees, 7.2 percent of Core STEM degrees, 6.5 percent of Healthcare degrees, and 7.2 percent of Other STEM degrees.

Figure 19. Racial and Ethnic Composition of Postsecondary Degrees Awarded in the 2011-2012 Academic Year.

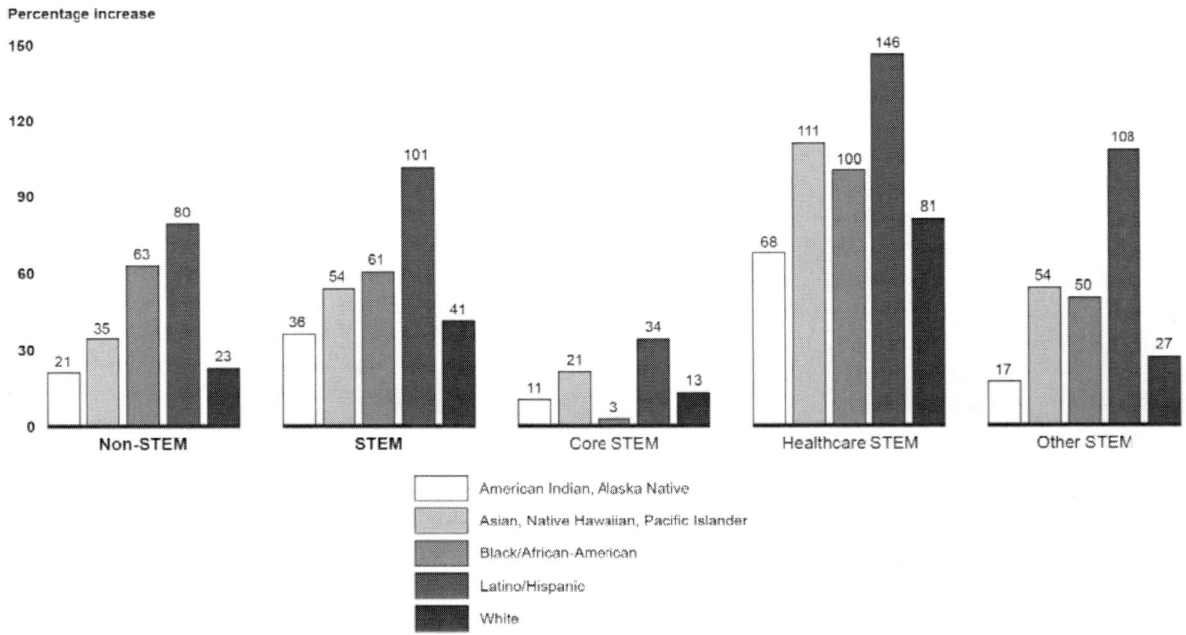

Source: GAO analysis of data from the Integrated Postsecondary Education Data System.

Figure 20. Percentage Increase in Postsecondary Degrees from the 2002-2003 to 2010-2011 Academic Years.

STEM Degrees by Race/Ethnicity

Among U.S. citizens and resident aliens, Asians and Pacific Islanders received a larger share of STEM degrees (7.1 percent), compared to their share of the non-STEM degrees (4.8 percent) (see figure 19). Other groups' share of STEM degrees was about the same as or less than their share of non-STEM degrees. Examining the data by STEM categories, however, African-Americans received a larger share of Healthcare degrees (15.1 percent), compared to their share of non-STEM degrees (12.6 percent).

Overall, STEM degrees awarded to Latino/Hispanic students increased more than other groups from the 2002-2003 to 2010-2011 academic years. STEM degrees have also increased at a higher rate among Asians and African-Americans, compared to whites. The increase among African-Americans was primarily in Healthcare and Other STEM fields (see figure 20).

APPENDIX III. EMPLOYMENT AND WAGE TRENDS IN STEM AND NON-STEM OCCUPATIONS

This appendix provides more detailed information about recent trends in STEM and non-STEM occupations.

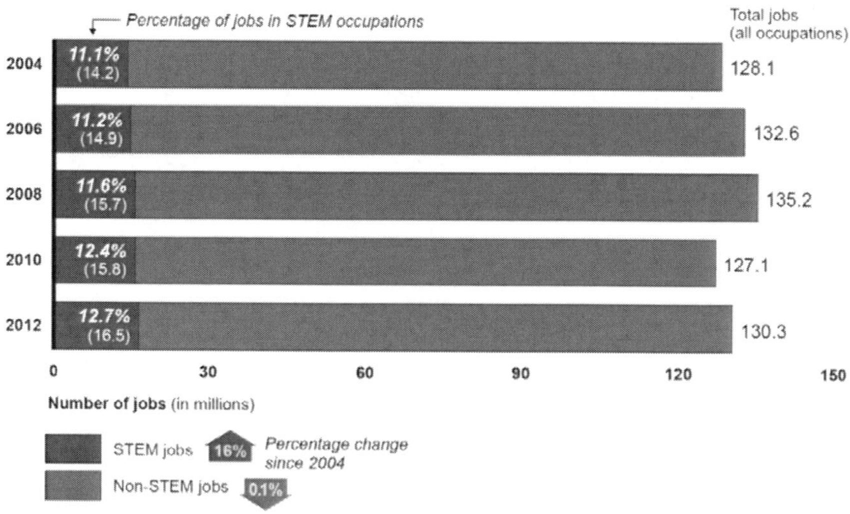

Source: GAO analysis of Occupational Employment Statistics data.

Note: Estimates of the number of jobs in STEM occupations shown in this figure have margins of error within plus or minus 0.5 percent of the estimate. Estimates of the number of jobs in non-STEM occupations have margins of error within plus or minus 0.2 percent of the estimate. With regard to the share of jobs in STEM occupations, the estimates have margins of error within plus or minus 0.1 percentage points. With regard to the percentage change in employment levels between May 2004 and May 2012, the estimates shown in this figure have margins of error within 0.3 percentage points. The difference between STEM and non-STEM occupations in the percentage change in employment levels between May 2004 and May 2012 is statistically significant at the 95 percent confidence level.

Figure 21. Employment in Science, Technology, Engineering, and Mathematics (STEM) and Non-STEM Occupations, 2004 to 2012.

Trends in STEM and Non-STEM Employment Levels

Figure 21 shows the number of jobs in STEM and non-STEM occupations from 2004 to 2012. While the overall number of jobs in STEM occupations increased throughout this time period, the number of jobs in non-STEM occupations declined during the recession.[57]

However, the trends vary by STEM categories. The number of jobs in Healthcare occupations steadily increased from 2004 to 2012, even during the recession (see figure 22). The number of jobs in Core STEM occupations, however, declined during the recession years, though it has increased since then.

Trends in STEM and Non-STEM Unemployment Rates

Unemployment rates in all three STEM categories have been relatively low from 2009 to 2012— about 5 percent or less—while rates in non-STEM occupations have been about 8 percent or higher. Unemployment rates in Healthcare occupations have been the lowest among the STEM categories, at less than 3 percent (see figure 23).

Trends in STEM Average Wages

Figure 24 shows trends in the average wage in the three STEM categories. The average wage was highest in core STEM occupations, but the greatest increase in wages occurred in healthcare occupations.

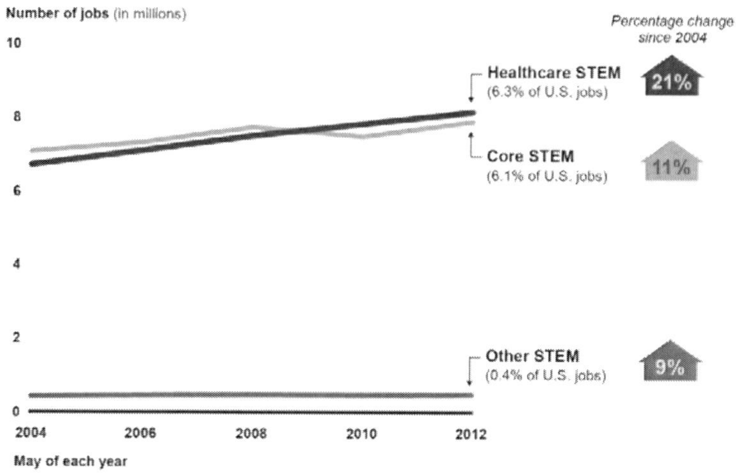

Source: GAO analysis of Occupational Employment Statistics data.

Note: Estimates of the number of jobs in Core STEM and Healthcare occupations have margins of error within plus or minus 0.7 percent. Estimates of the number of jobs in Other STEM occupations have margins of error within 2 percent of the estimate. With regard to the share of jobs in each category, the estimates have margins of error within plus or minus 0.03 percentage points. With regard to the percentage change in employment between May 2004 and May 2012, the estimates for Core STEM and Healthcare occupations shown in this figure have margins or error within plus or minus 0.4 percentage points. The estimate for the percentage change in jobs in Other STEM occupations between May 2004 and May 2012 has a margin of error within plus or minus 1.1 percentage points. The differences between the three STEM categories in the percentage change in employment levels between May 2004 and May 2012 are statistically significant at the 95 percent confidence level.

Figure 22. Employment in Science, Technology, Engineering, and Mathematics (STEM) Occupations, 2004-2012.

Education and Occupation

Thirty-eight percent of people with STEM bachelor's degree were working in STEM occupations in 2012, and the majority worked in non-STEM occupations. Figure 25 shows that much smaller percentages of workers with non-STEM bachelor's degrees or without a bachelor's degree worked in STEM occupations. However, they represented about half of workers in STEM occupations.

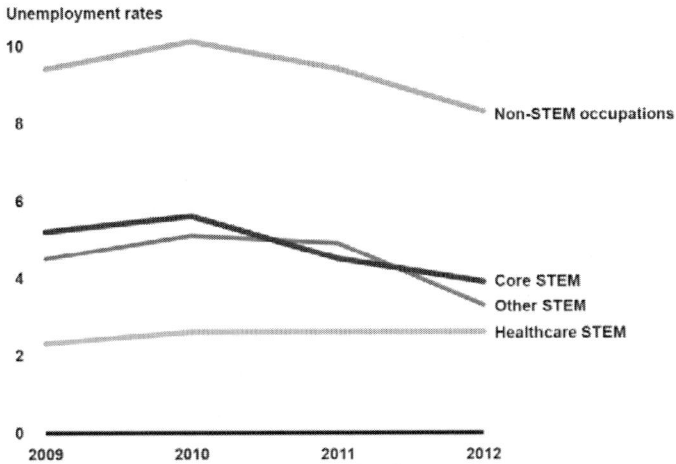

Source: GAO analysis of American Community Survey data.

Note: Estimates of unemployment rates for non-STEM, Core STEM and Healthcare STEM occupations shown in this figure have margins of error that are within plus or minus 0.2 percentage points. Estimates of unemployment rates for Other STEM occupations shown in this figure have margins of error that are within plus or minus 0.8 percentage points.

Figure 23. Unemployment Rates in Science, Technology, Engineering, and Mathematics (STEM) and Non-STEM Occupations, 2009 to 2012.

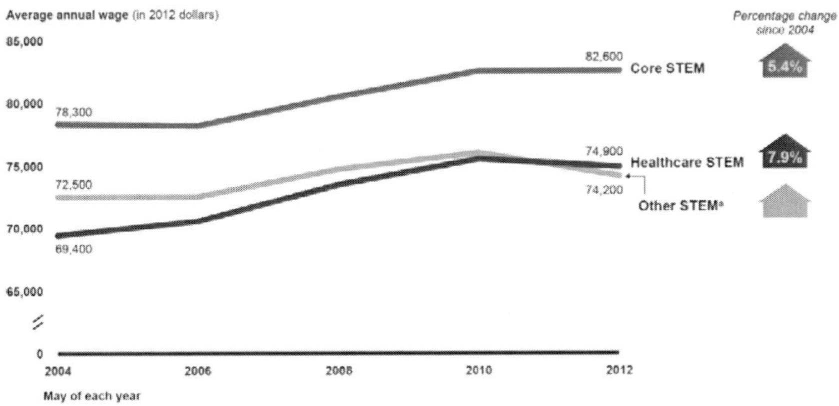

Source: GAO analysis of Occupational Employment Statistics data.

Note: Estimates shown in this figure of the average wage in Core STEM and Healthcare occupations have margins of errors within plus or minus 1.2 percent of the estimate. Estimates of the average wage in Other STEM occupations have margins of error within plus or minus 2.8 percent of the estimate. With regard to the percentage change in average wage between May 2004 and May 2012, the estimates for Core STEM and Healthcare occupations have margins of error within plus or minus 0.8 percentage points. The differences between the three STEM categories in the percentage change in the average wage between May 2004 and May 2012 are all statistically significant at the 95 percent confidence level.

[a] We do not report an estimate for this value because the margin of error at the 95 percent confidence level exceeds 30 percent of the estimate. The 95 percent confidence interval for the percentage change in average wage in Other STEM occupations is 0.5 to 4.0 percent.

Figure 24. Average Annual Wage in Science, Technology, Engineering, and Mathematics (STEM) Occupations, 2004 to 2012.

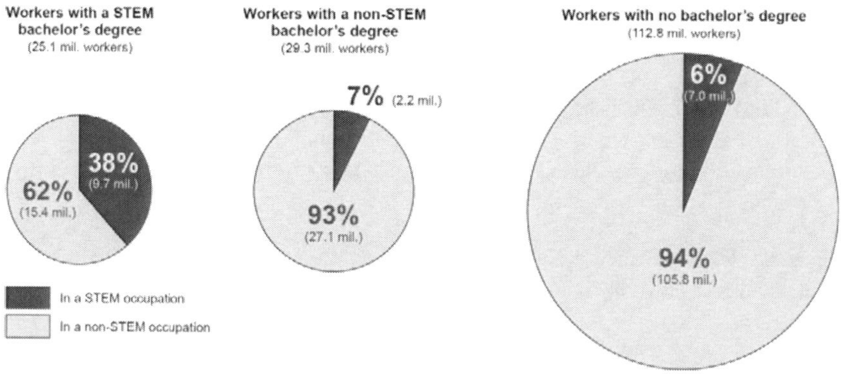

Source: GAO analysis of American Community Survey data.

Note: Estimates of the percentages in STEM and non-STEM occupations have margins of error that are within plus or minus 0.2 percentage points. Population estimates shown in this figure in parentheses have margins of error that are within plus or minus 0.02 percent of the estimate.

Figure 25. Percentage of Workers in Science, Technology, Engineering, and Mathematics (STEM) and Non-STEM Occupations by Educational Background, 2012.

Figure 26 shows the unemployment rates for the groups of workers shown in figure 25.

Figure 27 shows some non-STEM occupations with sizable populations of workers with STEM bachelor's degrees.

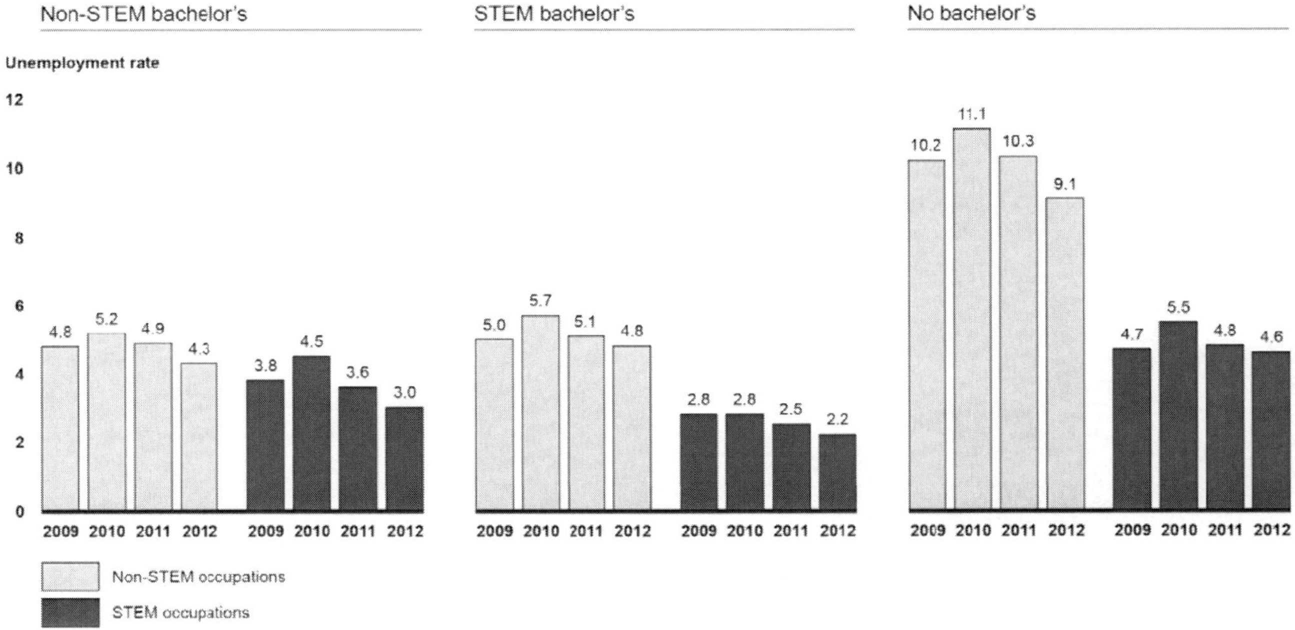

Source: GAO analysis of American Community Survey data.

Note: Estimates of the unemployment rate shown in this figure have margins of error that are within plus or minus 0.3 percentage points. Within each of the categories of STEM and non-STEM occupations, the differences between the non-STEM bachelor's, STEM bachelor's, and no bachelor's categories are statistically significant at the 95 percent confidence level in each of the years shown.

Figure 26. Unemployment Rates in Science, Technology, Engineering, and Mathematics (STEM) and Non-STEM Occupations, by Educational Background, 2009 to 2012.

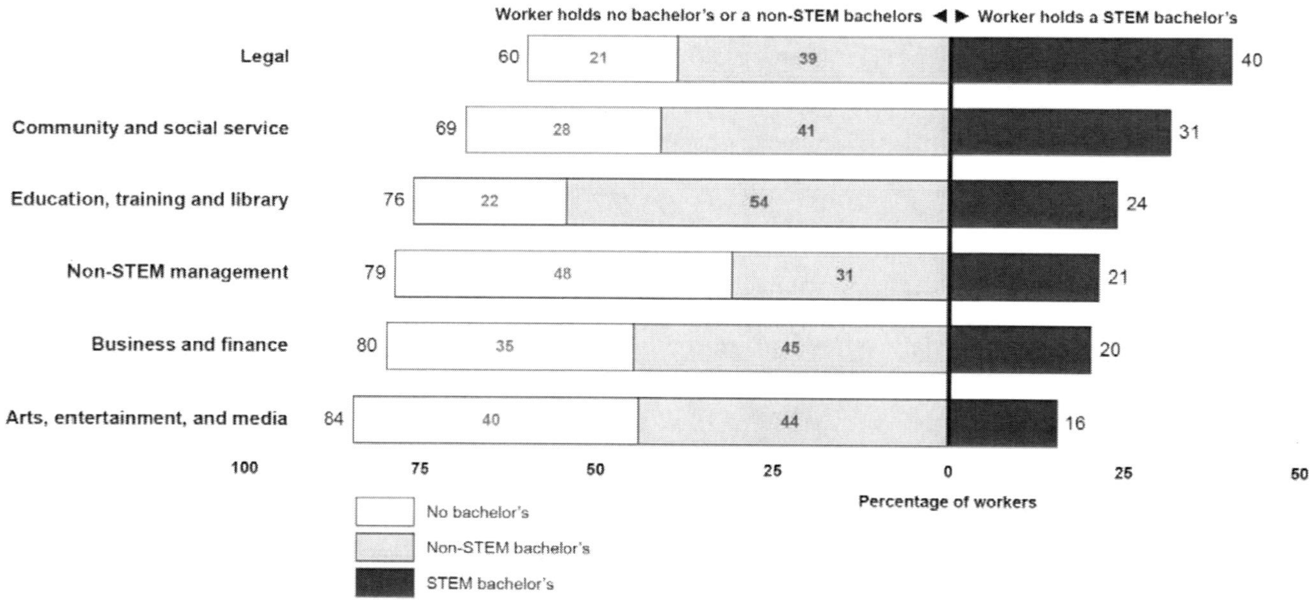

Source: GAO analysis of American Community Survey data.

Note: Estimates of the percentage of each occupational group in the education categories shown in this figure have margins of error that are within plus or minus 0.9 percentage points.

Figure 27. Educational Backgrounds of Workers Ages 22 or Older in Selected Non-Science, Technology, Engineering, and Mathematics (STEM) Occupations, 2012.

APPENDIX IV. SCIENCE, TECHNOLOGY, ENGINEERING, AND MATHEMATICS (STEM) PROGRAMS AND REPORTED FY 2012 OBLIGATIONS

Agency	Program	Fiscal Year 2012 STEM education program obligations[a]
NASA	Aeronautics Research Directorate - STEM Education activities	$3,300,000
	Aerospace Research and Career Development (ARCD) Program	$58,000,000
	Informal Education	$10,000,000
	Minority University Research and Education Project (MUREP)	$30,000,000
	Science Directorate - STEM Education activities	$41,000,000
	STEM Education and Accountability Projects — Higher Education	$21,000,000
	STEM Education and Accountability - Formal and Informal Education	$21,000,000
	Human Exploration and Operations (HEO) Mission Directorate -STEM Education Activities	$4,300,000
National Science Foundation	Advanced Technological Education (ATE)	$64,070,000
	Advancing Informal STEM Learning	$62,430,000
	Alliances for Graduate Education and the Professoriate (AGEP)	$7,840,000
	Discovery Research K-12 (DR-K12)	$99,570,000
	East Asia & Pacific Summer Institutes for U.S. Graduate Students (EAPSI)	$2,000,000
	Research in Engineering Education	$11,810,000
	Ethics Education in Science & Engineering (EESE)	$3,060,000
	CyberCorps(R): Scholarship for Service (SFS)	$44,980,000
	Graduate Research Fellowship (GRF) Program	$197,930,000
	Historically Black Colleges and Universities Undergraduate Program (HBCU-UP)	$31,850,000

Appendix IV. (Continued)

Agency	Program	Fiscal Year 2012 STEM education program obligations[a]
	Integrative Graduate Education and Research Traineeship (IGERT) Program	$65,430,000
	International Research Experiences for Students (IRES)	$100,000
	Louis Stokes Alliances for Minority Participation (LSAMP)	$45,480,000
	Math and Science Partnership Program (MSP)[b]	$57,070,000
	Nanotechnology Undergraduate Education in Engineering	$1,880,000
	Research Experiences for Teachers (RET) in Engineering and Computer Science	$7,870,000
	Research Experiences for Undergraduates (REU)	$79,550,000
	Research on Education and Learning (REAL)[c]	$54,160,000
	Robert Noyce Scholarship (Noyce) Program	$54,890,000
	Science, Technology, Engineering, and Mathematics Talent Expansion Program (STEP)	$25,300,000
	Transforming Undergrad Education in STEM (TUES)	$39,060,000
	Tribal Colleges and Universities Program (TCUP)	$13,390,000
Nuclear Regulatory Commission	Grants to Universities/Curriculum Development	$4,021,989
	Integrated University Program	$14,682,692
	Minority Serving Institutions Program	$2,800,000
Department of Agriculture		
Animal and Plant Health Inspection Service	AgDiscovery Program	$766,493
	1890 Institution Teaching, Research and Extension Capacity Building Grants Program	$22,456,532
National Institute of Food and Agriculture	Agriculture in the Classroom	$430,000

Agency	Program	Fiscal Year 2012 STEM education program obligations[a]
	Distance Education Grants Program for Institutions of Higher Education in Insular Areas	$7,500,000
	Food and Agricultural Sciences National Needs Graduate and Postdoctoral Fellowships Grants Program	$2,849,063
	Higher Education Challenge Grants Program	$4,500,000
	Higher Education Multicultural Scholars Program	$875,670
	Hispanic-Serving Institutions Education Grants Program	$9,000,000
	Resident Instruction Grants Program for Institutions of Higher Education in Insular Areas	$900,000
	Secondary Education, Two-Year Postsecondary Education and Agriculture in the K-12 Classroom Grants	$800,000
Office of the Assistant Secretary for Departmental Management	1890 National Scholars Program	$3,014,685
Department of Commerce		
National Institute of Standards and Technology	NIST Summer Institute for Middle School Science Teachers	$300,000
	Summer Undergraduate Research Fellowship (SURF) Program	$880,190
National Oceanic and Atmospheric Administration (NOAA)	Bay Watershed Education and Training (B-WET) Program	$5,490,619
	Environmental Literacy Grants	$2,626,990
	Dr. Nancy Foster Scholarship Program	$453,657
	Educational Partnership Program with Minority Serving Institutions	$12,500,000
	Ernest F. Hollings Undergraduate Scholarship Program	$4,959,273
	National Environmental Satellite, Data, and Information Service (NESDIS) Education	[d]
	National Marine Sanctuaries Education Program	[d]

(Continued)

Agency	Program	Fiscal Year 2012 STEM education program obligations[a]
	National Ocean Service (NOS) Education	[d]
	National Sea Grant College Program	$1,118,000
	National Weather Service (NWS) Education	[d]
	Teacher at Sea Program	$600,000
Department of Defense		
Air Force	National Defense Science and Engineering Graduate (NDSEG) Fellowship	$38,739,774
Army	Army Educational Outreach Program (AEOP)	$7,724,000
Navy	Navy STEM2Stern	$11,170,000
Department of Education	Developing Hispanic-Serving Institutions: STEM and Articulation Programs	$100,000,000
	Graduate Assistance in Areas of National Need	$30,873,072[e]
	Mathematics and Science Partnerships	$148,353,872[f]
	Minority Science and Engineering Improvement Program	$9,466,075
	Research in Special Education	$3,300,000
	Research, Development, and Dissemination	$31,200,000
	Strengthening Predominantly Black Institutions[g]	$15,000,000
	Upward Bound Math-Science	$44,141,410
Department of Energy[h]	Advanced Vehicle Competitions	$1,991,000[i]
	American Chemical Society Summer School in Nuclear and Radiochemistry	$561,000
	ASCR-ORNL Research Alliance in Math and Science	$250,000
	Community College Internships	$599,000[j]
	Computational Science Graduate Fellowship	$6,000,000
	Diversity in Science and Technology Advances National Clear Energy (DISTANCE)-Solar[k]	$365,000[l]
	Industrial Assessment Centers	$6,000,000
	Hampton University Graduate Studies	$46,000

Agency	Program	Fiscal Year 2012 STEM education program obligations[a]
	HBCU Mathematics, Science and Technology, Engineering and Research Workforce Development Program	$8,000,000
	Integrated University Program	$5,000,000
	Laboratory Equipment Donation Program	$124,000[m]
	Mickey Leland Energy	$655,000
	Minority Educational Institution Student Partnership Program (MEISPP)	$700,000
	National Science Bowl	$1,854,000[n]
	National Undergraduate Fellowship Program in Plasma Physics and Fusion Energy Sciences	$370,000
	Pan American Advanced Studies Institute	$200,000
	Plasma/Fusion Science Educator Programs	$209,000[o]
	QuarkNet	$610,000
	Science Undergraduate Laboratory Internships	$6,387,000[p]
	Solar Decathlon	$4,200,000[q]
	Summer Applied Geophysical Experience (SAGE)	$0[r]
	Visiting Faculty Program	$1,179,000[s]
Department of Health and Human Services		
Health Resources and Services Administration	Health Careers Opportunity Program	$14,779,000
National Institutes of Health	Bridges to the Baccalaureate Program	$8,200,000
	Bridges to the Doctorate	$3,600,000
	Cancer Education Grants Program (R25)	$12,473,029
	CCR/JHU Master of Science in Biotechnology Concentration in Molecular Targets and Drug Discovery Technologies	$301,400
	Center for Cancer Research Cancer Research Interns	$206,604
	Community College Summer Enrichment Program	$92,000

(Continued)

Agency	Program	Fiscal Year 2012 STEM education program obligations[a]
	Educational Programs for Demography and Population Science, Family Planning and Contraception, and Reproductive Research	$0[t]
	Graduate Program Partnerships	$11,121,000
	Initiative for Maximizing Student Development	$23,300,000
	Initiative to Maximize Research Education in Genomics	$1,336,000
	Intramural NIAID Research Opportunities	$935,429
	MARC U-STAR NRSA Program	$21,300,000
	Medical Infomatics Training Program	$6,074,705
	Medical Research Scholars Program	$1,100,000
	National Cancer Institute Cancer Education and Career Development Program (R25)	$18,285,877
	NIH Science Education Partnership Award (SEPA)	$18,616,000
	NIA MSTEM: Advancing Diversity in Aging Research (ADAR) through Undergraduate Education	$356,667
	NIAID Science Education Awards	$1,230,000
	Educational Programs for Population Research (R25)	$586,486
	NIDDK Education Program Grants	$322,529
	NIH Academy	$224,000
	NIH Summer Research Experience Programs	$1,951,274
	NIMH Mentoring Networks for Mental Health Research Education	$0[u]
	NIMH Research Education Programs for HIV/AIDS Research	$4,550,000
	NIMH Short Courses for Mental Health-Related Research Education	$0[v]
	NINR Summer Genetics Institute	$62,000
	Post-baccalaureate Intramural Research Training Award Program	$24,400,000
	Post-baccalaureate Research Education Program (PREP)	$7,700,000

Agency	Program	Fiscal Year 2012 STEM education program obligations[a]
	Research Education Grants for Statistical Training in the Genetics of Addiction	$683,058
	Research Supplements to Promote Diversity in Health-Related Research	$31,190,209
	RISE (Research Initiative for Scientific Enhancement)	$28,600,000
	Ruth L. Kirschstein National Research Service Award Institutional Research Training Grants (T32, T35)	$348,287,734
National Institutes of Health	Ruth L. Kirschstein NRSA for Individual Predoctoral Fellows, including Fellowships to Promote Diversity in Health-Related Research	$58,784,787
	Science Education Drug Abuse Partnership Award	$1,189,795
	Short Courses on Mathematical, Statistical, and Computational Tools for Studying Biological Systems	$1,200,000
	Short-Term Research Education Program to Increase Diversity in Health-Related Research	$4,706,540
	Student Intramural Research Training Award Program	$4,500,000
	Summer Institute for Training in Biostatistics	$0[w]
	Team-Based Design in Biomedical Engineering Education	$530,397
	Technical Intramural Research Training Award	$2,209,000
	Training in Computational Neuroscience: From Biology to Model and Back Again	$2,022,614
	Training in Neuroimaging: Integrating First Principles and Applications	$1,225,649
	Undergraduate Scholarship Program for Individuals from Disadvantaged Backgrounds	$2,400,000
Department of Homeland Security		
Science and Technology Directorate	Education - Career Development Grant Awards	$2,700,000
	HS-STEM Summer Internship Program	$350,000

(Continued)

Agency	Program	Fiscal Year 2012 STEM education program obligations[a]
	Minority Serving Institutions - Scientific Leadership Awards	$2,850,000
	Minority Serving Institutions - Summer Research Team	$600,000
Department of the Interior		
United States Geological Survey	EDMAP Component of the National Cooperative Geologic Mapping Program	$492,493
	National Association of Geoscience Teachers (NAGT)-USGS Cooperative Summer Field Training Program	$0
	Student Intern in Support of Native American Relations (SISNAR)	$0
Bureau of Land Management	Conservation and Land Management Internship Program	$2,500,000
National Park Service	Geoscientists-in-the-Parks Program	$596,090
Department of Transportation		
Federal Aviation Administration	Joint University Program	$450,000
	National Center of Excellence for Aviation Operations Research (NEXTOR)	$6,740,000
Federal Highway Administration	Garrett A. Morgan Technology and Transportation Education Program	$1,161,862
	National Summer Transportation Institute Program	$3,000,000
	Summer Transportation Internship Program for Diverse Groups	$1,100,000
Research and Innovative Technology Administration	University Transportation Centers Program	$72,000,000
Environmental Protection Agency	Cooperative Agreements for Training Cooperative Partnerships	$655,210
	Environmental Education Grants	$2,160,000
	EPA Marshall Scholars Program	$150,000
	Greater Research Opportunities Undergraduate Fellowship Program	$1,900,000

Agency	Program	Fiscal Year 2012 STEM education program obligations[a]
	National Environmental Education and Training Partnership	$2,259,500
	P3 Award: National Student Design Competition for Sustainability	$2,711,000
	President's Environmental Youth Awards	$1,200
	Science to Achieve Results Graduate Fellowship Program	$15,600,000
	Environmental Research Training Program	$1,391,069

Source: GAO survey of STEM education programs.

Note: Amounts obligated for each program for fiscal year 2012 were reported to us by agency officials in response to our survey. We did not independently verify this information.

[a] A few programs had zero obligations for fiscal year 2012. We determined that these programs still fit our definition of a STEM education program because they received federal funding and had not been terminated.

[b] The Math and Science Partnership program was consolidated with other programs into the Science, Technology, Engineering, Mathematics, including Computing Partnerships (STEM-C Partnerships) program in FY 2014.

[c] REAL combines three programs: Research and Evaluation on Education in Science and Engineering, Research in Disabilities Education, and Research on Gender in Science and Engineering.

[d] This program existed in fiscal year 2012, but did not respond to our survey.

[e] In response to our survey, Education reported $30,973,072 in obligations for the Graduate Assistance in Areas of National Need program, and we used that number for the analysis throughout the report. After our analysis was completed and the draft report was shared with Education, they reported that the actual obligations for fiscal year 2012 were $30,873,072. This represents a decrease of 0.32 percent from the reported program obligations and a decrease of 0.01 percent of total reported post-secondary STEM education program obligations. We determined that this change would not materially affect our overall results and findings and therefore we present our overall report analysis with the original survey submission. In addition, Education noted that funds for this program could be used to support non-STEM fields, such as area studies, foreign languages and literature, and educational evaluation also were allowable activities.

[f] In response to our survey, Education reported $148,000,000 in obligations for the Mathematics and Science Partnerships program, and we used that number for the analysis throughout the report. After our analysis was completed and the draft report was shared with Education, they reported that the actual obligations for

fiscal year 2012 were $148,353,872. This represents an increase of 0.24 percent in reported program obligations and a 0.05 percent increase in total reported K-12 STEM education program obligations. We determined that this change would not materially affect our overall results and findings and therefore we present our overall report analysis with the original survey submission.

[g] Obligations for Strengthening Predominantly Black Institutions were not exclusive to STEM activities. STEM is one of five allowable activities, and grantees can choose to focus their projects on any of these five activities.

[h] After our survey analysis was completed and the draft report was shared with Energy, officials reported changes to the fiscal year 2012 obligations for many of their programs. The changes to Energy's postsecondary programs summed to zero percent of total reported postsecondary obligations, and the changes to Energy's K-12 programs summed to zero percent of total reported K-12 obligations. We determined that these changes would not materially affect our overall results or findings. Individual changes are noted in table notes below.

[I] In response to our survey, Energy reported $1,992,000 in obligations for the Advanced Vehicle Competitions program, and we used that number for the analysis throughout the report. After our analysis was completed and the draft report was shared with Energy, they reported that the actual obligations for fiscal year 2012 were $1,991,000. This represents a 0.00 percent decrease in total reported postsecondary STEM education program obligations. We determined that this change would not materially affect our overall results and findings and therefore we present our overall report analysis with the original survey submission.

[j] In response to our survey, Energy reported $700,000 in obligations for the Community College Internships program, and we used that number for the analysis throughout the report. After our analysis was completed and the draft report was shared with Energy, they reported that the actual obligations for fiscal year 2012 were $599,000. This represents a 0.01 percent decrease in total reported postsecondary STEM education program obligations. We determined that this change would not materially affect our overall results and findings and therefore we present our overall report analysis with the original survey submission.

[k] In response to our survey, Energy reported $455,000 in obligations for the DISTANCE-Solar program, and we used that number for the analysis throughout the report. After our analysis was completed and the draft report was shared with Energy, they reported that the actual obligations for fiscal year 2012 were $365,000. This represents a 0.00 percent decrease in total reported postsecondary STEM education program obligations. We determined that this change would not materially affect our overall results and findings and therefore we present our overall report analysis with the original survey submission.

[l] In fiscal year 2012, the DISTANCE-Solar program was called the Minority University Research Associates program.

[m] In response to our survey, Energy reported $50,000 in obligations for the Laboratory Equipment Donation program, and we used that number for the analysis throughout the report. After our analysis was completed and the draft report was shared with Energy, they reported that the actual obligations for fiscal year 2012 were $124,000. This represents a 0.00 percent increase in total reported postsecondary STEM education program obligations. We determined that this change would not materially affect our overall results and findings and therefore we present our overall report analysis with the original survey submission.

[n] In response to our survey, Energy reported $2,800,000 in obligations for the National Science Bowl program, and we used that number for the analysis throughout the report. After our analysis was completed and the draft report was shared with Energy, they reported that the actual obligations for fiscal year 2012 were $1,854,000. This represents a 0.14 percent decrease in total reported K-12 STEM education program obligations. We determined that this change would not materially affect our overall results and findings and therefore we present our overall report analysis with the original survey submission.

[o] In response to our survey, Energy reported $774,000 in obligations for the Plasma/Fusion Science Educator Programs, and we used that number for the analysis throughout the report. After our analysis was completed and the draft report was shared with Energy, they reported that the actual obligations for fiscal year 2012 were $209,000. This represents a 0.08 percent decrease in total reported K-12 STEM education program obligations. We determined that this change would not materially affect our overall results and findings and therefore we present our overall report analysis with the original survey submission.

[p] In response to our survey, Energy reported $7,300,000 in obligations for the Science Undergraduate Laboratory Internships program, and we used that number for the analysis throughout the report.

After our analysis was completed and the draft report was shared with Energy, they reported that the actual obligations for fiscal year 2012 were $6,387,000. This represents a 0.05 percent decrease in total reported postsecondary STEM education program obligations. We determined that this change would not materially affect our overall results and findings and therefore we present our overall report analysis with the original survey submission.

[q] In response to our survey, Energy reported $2,250,000 in obligations for the Solar Decathlon program, and we used that number for the analysis throughout the report. After our analysis was completed and the draft report was shared with Energy, they reported that the actual obligations for fiscal year 2012 were $4,200,000. This represents a 0.10 percent increase in total reported postsecondary STEM education program obligations. We determined that this change would not materially affect our overall results and findings and therefore we present our overall report analysis with the original survey submission.

[r] In response to our survey, Energy reported $65,000 in obligations for the SAGE program, and we used that number for the analysis throughout the report. After our analysis was completed and the draft report was shared with Energy, they reported that the actual obligations for fiscal year 2012 were $0. This represents a 0.00 percent decrease in total reported postsecondary STEM education program obligations. We determined that this change would not materially affect our overall results and findings and therefore we present our overall report analysis with the original survey submission.

[s] In response to our survey, Energy reported $1,300,000 in obligations for the Visiting Faculty program, and we used that number for the analysis throughout the report. After our analysis was completed and the draft report was shared with Energy, they reported that the actual obligations for fiscal year 2012 were $1,179,000. This represents a 0.01 percent decrease in total reported postsecondary STEM education program obligations. We determined that this change would not materially affect our overall results and findings and therefore we present our overall report analysis with the original survey submission.

[t] No awards were made in response to program solicitations in fiscal year 2012.

[u] No funds were awarded in fiscal year 2012 because the first applications were due shortly before the end of the fiscal year.

[v] This program was not active in fiscal year 2012 and thus no funds were obligated in that fiscal year.

[w] Funds were obligated in fiscal year 2011 through fiscal year 2013 for grantees. Hence the obligation in fiscal year 2012 is $0.

End Notes

[1] See, for example, Hal Salzman, Daniel Kuehn, and B. Lindsay Lowell, *Guestworkers In The High-Skill U.S. Labor Market: An Analysis Of Supply, Employment, and Wage Trends*, Economic Policy Institute Briefing Paper #359 (Washington, D.C.: April 24, 2013).

[2] See, for example, Microsoft, *A National Talent Strategy: Ideas for Securing U.S. Competitiveness and Economic Growth.* (September 2012).

[3] See, for example, Anthony P. Carnevale, Nicole Smith, and Michelle Melton, *STEM: Science, Technology, Engineering, Mathematics*, Georgetown University Center on Education and the Workforce (October 20, 2011).

[4] See appendix I for more information about the four programs that did not respond to our survey. The survey also updated some of the descriptive data from our 2012 STEM report: GAO, *Science, Technology, Engineering, and Mathematics Education: Strategic Planning Needed to Better Manage Overlapping Programs across Multiple Agencies,* GAO-12-108 (Washington, D.C.: Jan. 20, 2012).

[5] Amounts obligated for each program for fiscal year 2012 were reported to us by agency officials and we did not independently verify this information. An obligation is a definite commitment that creates a legal liability of the government for the payment of goods and services ordered or received, or a legal duty on the part of the United States that could mature into a legal liability. Payment on these obligations may be made immediately or in the future. An agency incurs an obligation, for example, when it places an order, signs a

contract, awards a grant, or purchases a service. See GAO, *A Glossary of Terms Used in the Federal Budget Process,* GAO-05-734SP (Washington, D.C.: Sept. 2005).

[6] Informal education programs support activities provided by a variety of organizations that offer students learning opportunities outside of formal schooling through contests, science fairs, summer programs, and other means. Outreach programs targeted to the general public (either adults or children) are not included.

[7] GAO-12-108. In our 2012 report, we considered federal STEM education programs funded in fiscal year 2010. In the current report, we consider those funded in fiscal year 2012.

[8] The current administration defines a STEM education investment (it does not use the word "program") as a federally funded STEM education activity that has a dedicated budget of $300,000 or above and staff to manage the budget. Our definition does not have a stated budget minimum. See Committee on STEM Education, National Science and Technology Council, *Federal Science, Technology, Engineering, and Mathematics (STEM) Education 5-Year Strategic Plan* (Washington, D.C.: May 2013).

[9] We based our categories on a categorization of STEM put forth by the Standard Occupational Classification Policy Committee presented in *Options for Defining STEM (Science, Technology, Engineering, and Mathematics) Occupations Under the Standard 2010 Occupation Classification System*, Standard Occupational Classification Policy Committee Recommendation to the Office of Management and Budget (August 2012). In 2011, the Standard Occupational Classification Policy Committee, a federal inter-agency committee responsible for recommending updates to the classification system used in occupational data, developed several options for defining STEM occupations. These included a categorization into the following four areas: (1) life and physical science, engineering, mathematics, and information technology occupations, (2) social science occupations, (3) architecture occupations, and (4) health occupations.

[10] GAO-12-108.

[11] Committee on STEM Education, National Science and Technology Council, *Coordinating Federal Science, Technology, Engineering, And Mathematics (STEM) Education Investments: Progress Report* (Washington, D.C.: February 2012); Committee on STEM Education, National Science and Technology Council, *Federal Science, Technology, Engineering, and Mathematics (STEM) Education 5-Year Strategic Plan. (*Washington, D.C.: May 2013).

[12] Office of Management and Budget, *Budget of the United States Government, Fiscal Year 2014* (Washington, D.C.: April 2013).

[13] Office of Management and Budget, *Budget of the United States Government, Fiscal Year 2015* (Washington, D.C.: March 2014).

[14] We list these priority areas as stated in the 5-year Strategic Plan.

[15] In this report, postsecondary degrees refers to and includes associate's and other degrees awarded below the bachelor's level, bachelor's, master's, postbaccalaureate and post-master's certificates, doctorate, and professional degrees. We include both degrees awarded for first and second majors in our analysis. Our results represent the number of degrees awarded, not the number of individuals who obtained degrees.

[16] When degrees awarded to non-resident aliens are excluded, degrees awarded in core STEM fields increased 18 percent between the 2002-2003 and 2011-2012 academic years, 100 percent in healthcare fields, and 44 percent in other STEM fields. Overall, degrees awarded to citizens and resident aliens increased 56 percent in STEM fields in this time period and 37 percent in non-STEM fields. Degrees awarded to non-resident alien students comprised 4 percent of all degrees awarded in the 2011-2012 school year. However, non-resident alien students were more heavily concentrated in core STEM fields, particularly at the graduate level—23 percent of non-resident alien degrees were in core STEM fields at the master's and the doctorate or professional level, compared to 2 percent of degrees awarded to U.S. citizens and residents. See appendix II for further information on STEM degrees awarded to various demographic groups.

[17] Specifically, degrees at the below-bachelor's, bachelor's, and master's level in computer science/IT fields declined for 3 to 5 years from the 2002-2003 academic year and then increased. Computer science/IT degrees at the post-bachelor's certificate level was fairly stable (at about 800 to 900 degrees awarded) from 2002-2003 to 2009-2010 period but increased to about 1,200 degrees awarded in the 2010-2011 and 2011-2012 academic years. Computer science/IT degrees at the doctorate level steadily increased through the past decade, from about 800 degrees awarded in 2002-2003 to about 1,700 in the 2011-2012 academic year. Overall trends in computer science/IT degrees awarded were similar for non-resident alien students and citizens and residents—computer science/IT degrees declined 17 percent among non-resident aliens and 19 percent among citizens and residents from the 2002-2003 to 2011-2012 academic years. Citizens and residents received the large majority of computer science/IT degrees awarded at the less than bachelor's (99 percent), bachelor's (95 percent), and post-bachelor's certificate levels (84 percent) in the 2011-2012 academic year. At the graduate levels, sizable percentages of computer science/IT degrees were awarded to non-resident alien students (44 percent of master's degrees and 51 percent of doctorate degrees).

[18] Stuart Zweben, *Computing Degrees and Enrollment Trends from the 2009-2010 CRA Taulbee Survey*, Computing Research Association (Washington, D.C.).

[19] Growth in degrees for the Core STEM field of engineering or science technician has been relatively slow. These degrees prepare students for jobs like industrial production technicians, telecommunications technicians, solar energy technicians, nuclear and industrial radiologic technicians. Degrees in these fields, which are largely at the below-bachelor's level, increased only 10 percent in the past decade (see figure 5). Similar to computer science/IT degrees, the number of technician degrees awarded declined from the 2002-2003 to 2007-2008 academic years, but has been increasing since then.

[20] We do not present unemployment rates for STEM sales occupations because the standard errors exceeded 30 percent of the estimates at the 95 percent confidence level.

[21] See, for example, Richard B. Freeman, "A Cobweb Model of the Supply and Starting Salary of New Engineers, *Industrial and Labor Relations Review*, 29: 2 (January 1976), 236-248; Jaewoo Ryoo and Sherwin Rosen, "The Engineering Labor Market," *Journal of Political Economy*, 112: 1 pt. 2 (February 2004), 110-140.

[22] Supply and demand for STEM workers in the United States are also affected by global factors. The supply of STEM workers is affected by the number of foreign workers who relocate to the United States and work in STEM occupations. Demand for STEM workers in the United States is affected by global demand for American firms' products and services, as well as the presence of STEM workforces in other countries and the extent to which United States firms relocate activities or operations overseas to access those workers. For further information, see GAO, *H-1B Visa Program: Reforms Are Needed to Minimize the Risks and Costs of Current Program*, GAO-11-26 (Washington, D.C.: Jan. 14, 2011) and *Offshoring of Services: An Overview of the Issues*, GAO-06-5 (Washington, D.C.: Nov. 27, 2005).

[23] It is possible that these workers might have received another degree in the field in which they were working, but the American Community Survey data do not show this, since this survey only captures information on the field of study for degrees at the bachelor's level.

[24] These programs represented 95 percent of fiscal year 2012 obligations by federal postsecondary STEM education programs that responded to our survey.

[25] For purposes of this report, we consider federal postsecondary STEM education programs to also include programs with both a postsecondary and K-12 component.

[26] We created a list of possible workforce needs using input from experts, program officials, and grantees, and asked federal STEM education programs to indicate whether each possible workforce need was a stated program objective, a potential benefit of the program, or neither.

[27] We analyzed the proportions of both programs and obligations dedicated to a particular objective. See appendix I for a detailed explanation. Amounts obligated for each program

for fiscal year 2012 were reported to us by agency officials in response to our survey. We did not independently verify this information.

[28] While the number of programs with stated objectives and potential benefits related to specific occupations were roughly equal, much more money was obligated by the programs with potential benefits ($989 million, or 53 percent of postsecondary STEM obligations), compared to those with stated objectives (23 percent of obligations, or $429 million). This may indicate that the programs with stated objectives had smaller budgets, on average.

[29] Twenty-four grant-making programs, with obligations of $291 million, had a stated objective to focus on specific STEM occupations. Of these, 13, with $214 million in obligations, gave preference to grant applicants that focused on the same occupations.

[30] Ninety-nine federal postsecondary STEM programs ($1.4 billion in obligations) focused on specific occupations as either a stated program objective or potential benefit of their work. Of these, 55 programs ($349 million in obligations) identified occupations based on market demand. We defined high market demand to include STEM occupations with many job openings now, many predicted job openings in the future, or a shortage of qualified workers (e.g., there are not enough qualified workers available to fill job openings).

[31] Ninety-nine federal postsecondary STEM programs ($1.4 billion in obligations) focused on specific occupations as either a stated program objective or potential benefit of their work. Of these, 84 programs ($919 million in obligations) chose occupations related to their agency's mission, including 51 programs that considered both market demand and mission.

[32] Forty-seven programs indicated on our survey that they had a stated program objective to increase the numbers of minority, disadvantaged, or under-represented groups in the STEM workforce. Of these, 17 did not respond to questions asking which specific groups they focused on. The fairly high number of nonrespondents suggests that this information should be interpreted with caution.

[33] Eighty postsecondary programs, with $1.396 billion in obligations in fiscal year 2012, indicated that they made grants for the purposes of STEM education. Of these, 39 programs, with $1.07 billion in obligations, gave preference to applicants that focused on minority, disadvantaged, or under-represented groups.

[34] National Institutes of Health, *Biomedical Research Workforce Working Group Report: A Working Group of the Advisory Committee to the Director* (June 14, 2012). The National Institutes of Health has many programs that fund biomedical students. According to the report, the majority of biomedical graduate students receive National Institutes of Health support at some point in their graduate studies through training grants, fellowships or research project grants. It also found that although the agency primarily trained graduate students for careers in academic research, less than half of domestically trained biomedical PhD graduates in 2008 went into a career in academia, with many going into research or non research careers in government and industry. As a result, the report recommended that the National Institutes of Health create a program providing additional training and career development experiences to equip students for various career options.

[35] We also found that STEM programs varied in their ability to track reliable output measures, such as the number of students or teachers directly served by their program. We recommended that the National Science and Technology Council develop a monitoring framework to ensure that agencies collect data and report on the goals in the Strategic Plan. Work on this framework has begun, but our recommendation remains open because the framework has not been completed. See GAO-12-108.

[36] For purposes of this report, we limited our category of federal K-12 STEM education programs to include only those STEM education programs that primarily serve students and teachers at the K-12 level. The category is not meant to be inclusive of all programs with a K-12 STEM education component.

[37] We analyzed the proportions of both programs and obligations for fiscal year 2012. Amounts obligated for each program for fiscal year 2012 were reported to us by agency officials in response to our survey. We did not independently verify this information.

[38] In addition to the 13 K-12 programs, 7 postsecondary programs noted that preparing students for postsecondary STEM education is a stated program objective.

[39] These programs are required to provide instruction in math and science—as well as laboratory experience, academic counseling, information on and assistance with applying for student aid, and mentorships opportunities.

[40] In addition to the 18 K-12 programs, 4 postsecondary programs noted that improving the ability of K-12 teachers to teach STEM content is a stated program objective.

[41] NSF administers the Math and Science Partnership program, while Education administers the Mathematics and Science Partnerships program.

[42] Education releases an annual performance report that aggregates results from all Upward Bound Math-Science programs and presents program-wide outcomes.

[43] GAO, *Science, Technology, Engineering, and Mathematics Education: Strategic Planning Needed to Better Manage Multiple Programs Across Multiple Agencies*, GAO-12-108 (Washington, D.C.: Jan. 20, 2012).

[44] 20 U.S.C. §1094(a)(17).

[45] The OES survey does not collect data from self-employed persons. As a result, our estimates from the OES data do not include data from self-employed persons.

[46] *Options for Defining STEM (Science, Technology, Engineering, and Mathematics Occupations Under the 2010 Standard Occupational Classification (SOC) System,* SOC Policy Committee Recommendation to the Office of Management and Budget (OMB) (August 2012).

[47] Three occupations were classified by the SOC Policy Committee in multiple categories: architectural and engineering managers, architectural and civil drafters, and life, physical, and social science technicians, all other. We classified these occupations as Core STEM in our analysis.

[48] The 2010 OES also used some temporary SOC codes for occupations where the estimates were based on some surveys that used the 2000 SOC and some surveys that used the 2010 SOC. We reviewed each of these temporary codes and classified them as STEM or non-STEM based on the SOC Policy Committee's *Options for Defining STEM Occupations*.

[49] GAO-12-108. In that report, we inventoried federal STEM education program funded in fiscal year 2010. In the current report, we inventoried those funded in fiscal year 2012.

[50] After we had deployed our survey, program officials recommended that we exclude five programs from our review. After speaking with officials and reviewing program information, we determined that these five programs should be excluded from our list and should not complete the survey.

[51] We chose these programs because they were among the largest federal STEM education programs, collectively accounting for 54 percent of the fiscal year 2012 STEM education obligations reported by respondents to our survey.

[52] Four programs did not respond to our survey, all from the National Oceanic and Atmospheric Administration. They were the education programs housed in the National Marine Sanctuaries; National Environmental Satellite, Data, and Information Service; National Ocean Service; and National Weather Service.

[53] These agencies are the Environmental Protection Agency, the National Aeronautics and Space Administration, the National Science Foundation, the Nuclear Regulatory Commission, and the U.S. Departments of Agriculture, Commerce, Defense, Education, Energy, Health and Human Services, Homeland Security, Interior, and Transportation.

[54] These agencies are the Environmental Protection Agency, the National Aeronautics and Space Administration, the National Science Foundation, and the U.S. Departments of Agriculture, Commerce, Defense, Education, Energy, Health and Human Services, and Transportation.

[55] Men received the majority of Core STEM degrees at all levels (less than bachelor's, bachelor's, post-bachelor's certificates, master's, and doctorate/professional). Women received the majority of Healthcare degrees at all levels. At the doctorate/professional level, women's

share of Healthcare STEM degrees awarded has increased from 51 percent in the 2002-2003 academic year to 58 percent in the 2011-2012 academic year.

[56] Men received the majority of postsecondary degrees in these fields at all degree levels. Women received the majority of life sciences degrees at all degree levels.

[57] The recent recession officially began in December 2007 and ended in June 2009.

INDEX

D

E

F

G

H

I

J

K

L

M

N

O

P

Q

R

S

T

U

V

W